Optoelectronic Line Transmission

Optoelectronic Line Transmission

An introduction to fibre optics

Raymond L. Tricker

BH NEWNES

Butterworth-Heinemann Ltd
Linacre House, Jordan Hill, Oxford OX2 8DP

PART OF REED INTERNATIONAL BOOKS

OXFORD LONDON BOSTON
MUNICH NEW DELHI SINGAPORE SYDNEY
TOKYO TORONTO WELLINGTON

First published 1989
Reprinted 1992

British Library Cataloguing in Publication Data
Tricker, Raymond L.
Optoelectronic line transmission
1. Fibre – optic data. Transmission equipment
I. Title
621.38'028

ISBN 0 7506 0826 9

Printed and bound in Great Britain by
Redwood Press Limited, Melksham, Wiltshire

Contents

Preface

During the last few years the use of optical fibre has become increasingly popular in both commercial and military environments and although **optoelectronics** is based upon a comparatively simple technology, it is nevertheless essential for today's engineers to be aware of the basic fundamentals and capabilities of this modern technique.

In order to understand new electronic engineering principles, however, it is usually necessary for the student to have a thorough technical foundation and for the author to explain mathematical theories and progressively develop formulae etc. to enable the reader to continue his studies with a complete understanding of the fundamentals surrounding this new technology.

Although an understanding of the mathematical principles applicable to optoelectronics is of course very necessary, it is the aim of this book to provide a basic introduction as well as a background reference manual to fibre optic transmission for practising electronic and telecommunications engineers, technicians and students who would like to get to grips with this new area of communications technology.

Acknowledgements

I would like to thank my wife Claire for giving me the inspiration to work for a degree with the Pacific Western University. This resulted in a thesis which became the basis of this book. I would like also to acknowledge the help given to me by the Editor of *Electronic Technology* (Mr I. R. G. Channing) and pay a special thank you to Carol Crossfield for so painstakingly proofreading the manuscript, supplying many helpful comments but most of all, for saying that she would even have a copy when it was printed!

1
Introduction

BACKGROUND

For over a century, telegraph and telephone systems have been interconnected via copper cables and radio links and these have formed the basis of all suburban, urban and national telecommunication networks.

Although line transmission technology has improved over the years, it was not until the last decade or so when revolutionary innovations caused by the development of **semiconductors, digitalisation** and **optoelectronics,** dramatically altered the method, technology and cost effectiveness of communication systems.

Of course, the transmission of information by visual means is not an entirely new concept, as man has been sending messages optically for centuries. For instance, in the sixth century BC, Aeschylus used a long chain of bonfires from Asia minor to Argos to pass on the news of Troy's downfall. In the second century BC, Polybius developed a two digit, five level code by which the whole of the Greek alphabet could be transmitted [1].

It is perhaps interesting to note that it is this last method that is generally recognised as being the first example of **optoelectronic signalling.**

The Chinese, when threatened by the Tartars, lit signal fires along the boarders of their empire, increasing brightness by artificial means to penetrate fog and rain. Gengis Khan (1115–1227) controlled and checked his horsemen by flag telegraphy. There are many examples from this particular period.

During the next few decades a number of experiments were carried out utilising various other visual methods. For example, in 1633 the Marquis of Worcester developed long distance communication systems using black and white signs, and in the papers of John Norris of Hughenden Manor in Buckinghamshire [2], there is an enigmatic comment: '3 June 1778. Did this day Heliograph Intelligence from Dr Franklin in Paris to Wycombe'. Norris built a 30 m tower on a hill at Camberley, Surrey, from the top of which he used to signal and place bets by heliograph with Lord le Despencer at West Wycombe.

Turning the pages of history once again, it was in 1795 that Lord George Murray was commissioned to supervise the installation of a chain of shutter stations stretching from Plymouth to London [2].

These stations formed the Admiralty Shutter Telegraph system (Figure 1.1) and consisted of six, 3-foot square shutters in a frame measuring 30 foot by 20 foot.

By careful positioning of the shutters, a total of 63 changes were possible: sufficient to cater for the whole of the alphabet, the ten numerals, as well as a number of selected phrases.

The stations were normally manned by three civilians (one to operate the shutters, the other two to act as telescope observers) and it is recorded [3] that an average message passed from London to Portsmouth in fifteen minutes whereas a pre-arranged signal could be communicated between London and Portsmouth and acknowledged back in two minutes and between London and Plymouth, via the Blandford Racecourse station, in three minutes.

Across the channel and a few years previous to the Admiralty scheme (i.e. 1792), Claude Chappé had succeeded in passing a message between Paris and Strasburg (a distance of over 400 km) using movable signal elements. Chappé's device (the T-type telegraph – Figure 1.2) consisted of two side wings mounted at right angles to a main wing that was positioned at the top of a mast. By changing the position of the main wing with respect to the mast, varying the attitude of the side wings to the main wing by 45 degree steps, 196 different and easily distinguishable signals could be given. These signals were

(a)

Figure 1.1
The Admiralty shutter telegraph system (a) artist's impression of a relay station (courtesy of Royal Signals Museum, Blandford) (b) map showing relay station sites between Plymouth and London

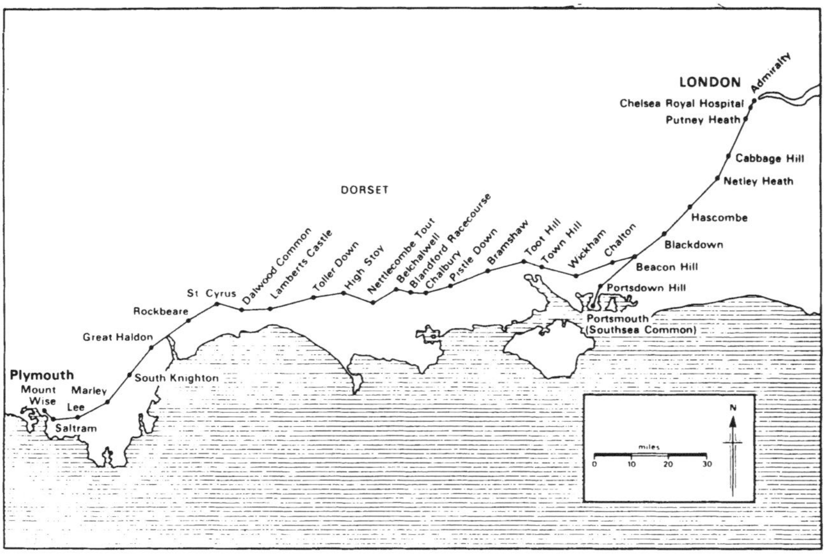

(b)

Figure 1.2 The Chappé radiated telegraph, near Waterloo (courtesy of Royal Signals Museum, Blandford)

observed from nearby stations using telescopes and then passed on to the next relay station in a similar manner [4].

Napoleon, who quickly recognised the importance of optical signalling, established telegraph stations in the countries in which he was campaigning and equipped his armies with optical telegraph equipment (**telegraph ambutantes**), similar to Col John Macdonald's **Anthropological Telegraph** illustrated in Figure 1.3.

In Prussia, General Freiherr von Muffling carried out a number of trials during the period 1819 to 1830 utilising optical telegraph systems. The success of these studies led to the construction of the first telegraph line between Berlin and Potsdam. During this time various **heliograph** and **signal lamp systems** were also developed enabling signalling over distances of around 100 kilometres.

Although heliograph communications were, by that

Figure 1.3 Colonel John Macdonald's Anthropological Telegraph (courtesy of Royal Signals Museum, Blandford)

time, well advanced it is interesting to note that Henry Coxwell, aware that heliography could only be used on cloudless days, developed **balloon signalling** in the 1850s [5].

Figure 1.4 represents Coxwell's method of adapting semaphore arms of various shapes and symbols to convey, according to preconcerted arrangement, any required information to those on the ground.

A few years later (1870) John Tyndall gave a presentation to the Royal Society showing that light would follow a curved jet of water [4] [6]. But it was not until the 1880s that Graham Bell (in conjunction with Sununer Tainter) successfully modulated a beam of sunlight using a diaphragm mirror [7]. The action of the **photophone** is very simple (Figure 1.5a). Rays of sunlight (or powerful light source) are reflected by a flat mirror (h) into a system of lenses (a and b) that form a beam of light onto a silver coated glass plate (H). A person speaking into the mouthpiece of the transmitter (see also Figure 1.5b) causes the silver coated plate to vibrate which in turn causes the reflecting surface of the plate to change shape. This effects the strength of the light rays which are compressed by a second set of lenses (c) and then reflected to the receiver.

At the distinct end (see also Figure 1.5c) the light rays are received by a silver-coated copper parabolic reflecting mirror (P), and by using a sensitive selenium-photoresistor (S) (comprising a cylinder of copper plates, separated by strips of mica and filled with selenium – whose resistance varies according to the quantity of light that strikes it) connected to a battery and a telephone instrument (t), the system is able to reproduce the speaker's voice.

All of these methods were of course affected by the visual transparency of the medium through which they were being propagated (ie, air) and fog, rain, storms, cloud and snow were some of the problems to be contended with.

In an attempt to overcome these problems, the possibil-

Figure 1.4
Henry Coxwell's balloon signalling (from engravings published in the *London Illustrated News*, 8 October 1879 – courtesy Royal Signals Museum, Blandford)

ity of using a **dielectric rod** as a **waveguide** was studied during the early 1900s. Although a paper was written about it in 1910, it was not until 1966 that Charles Kao and George Hockham (two scientists working for Standard Telecommunications Laboratories in Harlow) patented the principle of information transmission though a transparent dielectric medium, (ie, glass fibre). It was from these patents that the use of glass fibres actually became a viable proposition.

Since the development of **optical fibres**, which in most

cases are no thicker than a human hair (Plate 1.1), more and more communication systems are now using fibres to pass optical signals over long distances with very little loss. (Compared to the losses experienced by electrical signals over copper lines.) Following the development of **laser diodes**, the use of optoelectronics has now become the *preferred* method for cable communication.

Basic principles of optical line transmission

Some **crystals**, for example **Gallium Arsenide** (GaAs) emit light at a wavelength very close to visible light when an

Figure 1.5
Bell and Tainter's photophone (a) component parts (b) the transmitter (c) the receiver (extracted from *Victorian Inventions* by Leonardo de Vries)

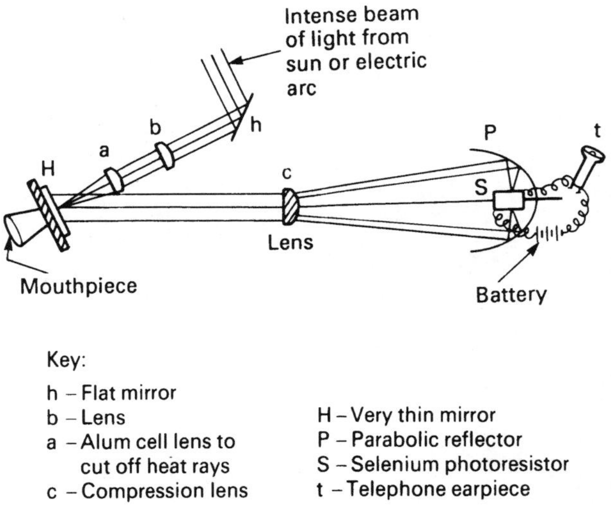

(a)

(b)

(c)

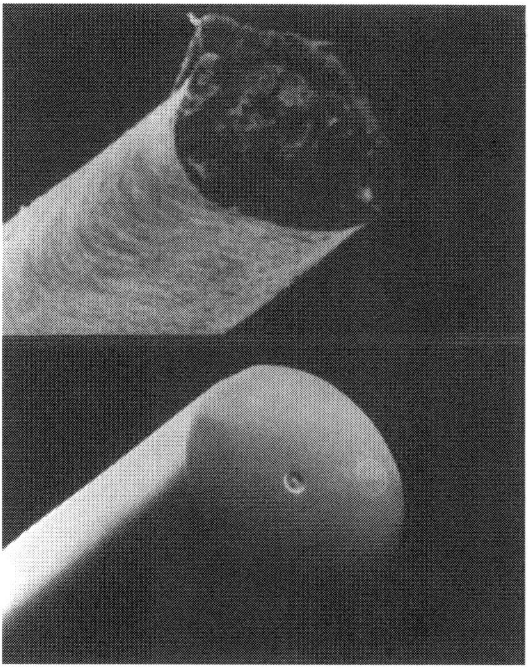

Plate 1.1
A human hair (top) and an optical fibre (bottom) (courtesy Siemens Aktiengesellschaft)

electric current is passed through them. Figure 1.6 shows the frequency and wavelength of common signalling methods as an electromagnetic spectrum.

This light emission is produced by the release of energy (**photons**) from the atoms of a material and is caused when the atoms are excited by heat, chemical reaction or some other means. Utilising this principle it is possible to communicate (for example by telephone) over an optical link, as shown in Figure 1.7.

The microphone in the telephone set converts speech, (ie, sound waves), into electrical energy. A **multiplexer** is used to combine this with other similar **information channels** together to form an **information group**. This is **coded** (if required) and amplified prior to being converted into light signals by an **optoelectric crystal transducer** (usually a light emitting diode or laser diode) and transmitted over an optical fibre waveguide. This fibre, of approximately 0.1 mm in diameter, is drawn from high

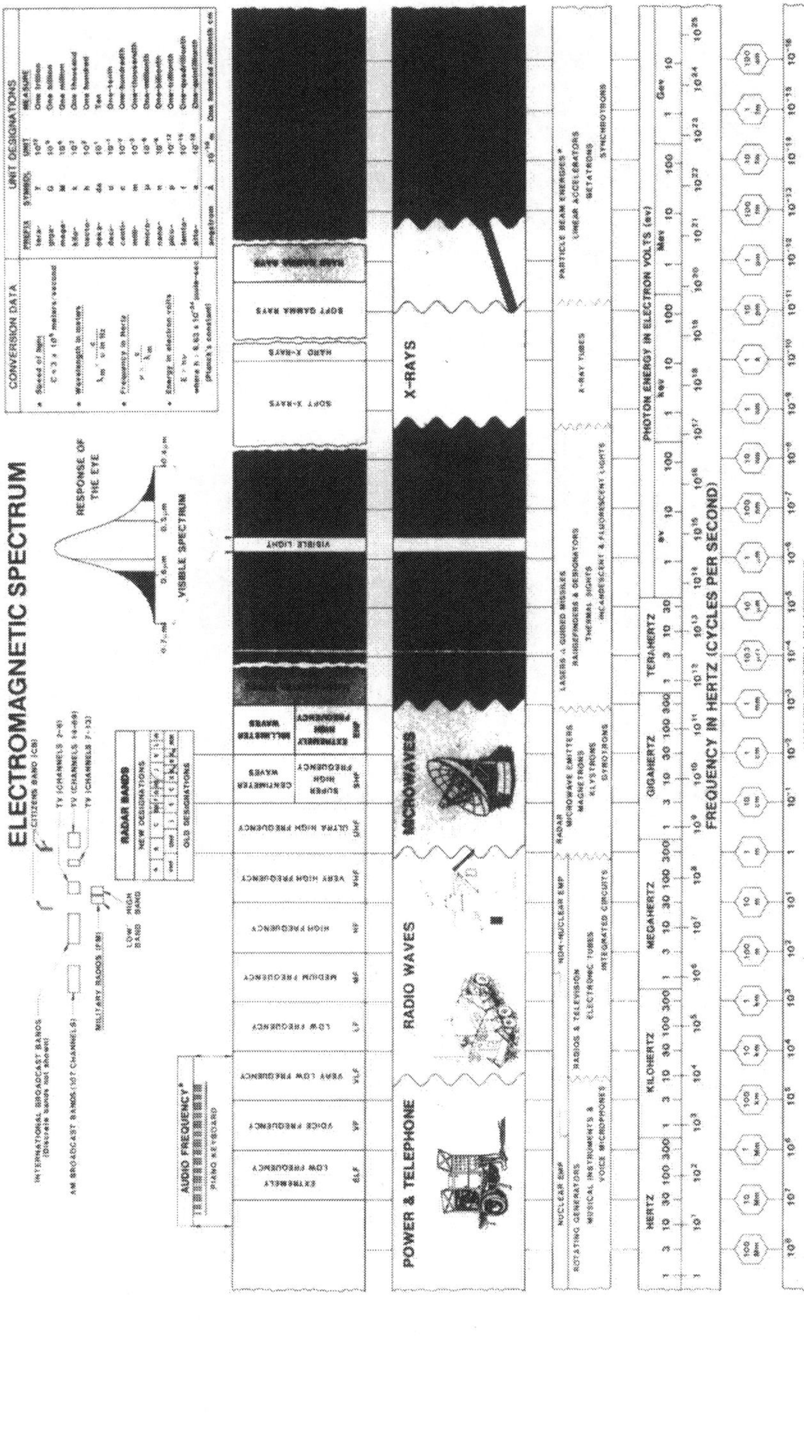

*Not Electromagnetic Waves

Figure 1.6 Electromagnetic spectrum, showing wavelengths and frequencies of common signals methods courtesy of BDM International Inco

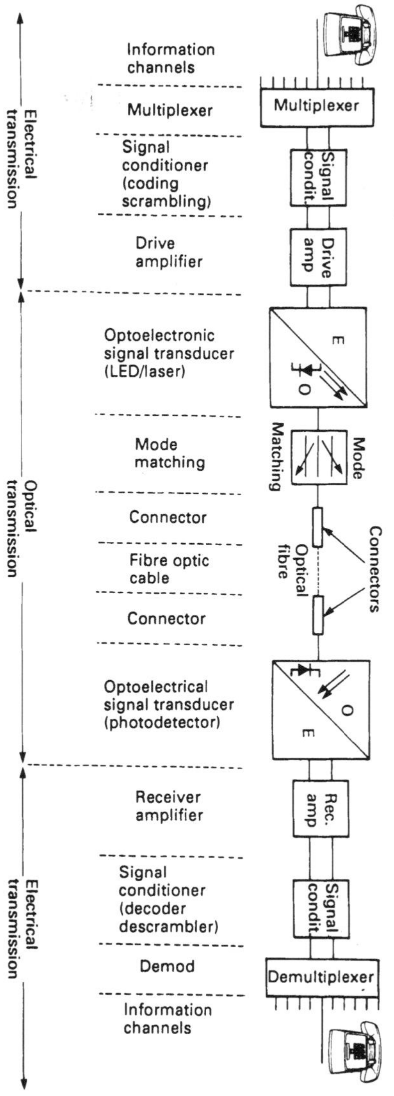

Figure 1.7 Basic optoelectronic communication system

purity synthetic silica glass, and may be several kilometres in length.

At the far end of the optical link the light signals are reconverted into electrical energy by another type of optoelectronic signal transducer called a **photodetector**: amplified, decoded, split back into their individual speech channels and then converted into sound waves, that are audible to the subscriber, through the use of **electro-acoustic transducer** – i.e. telephone earpiece.

Avantages of optical fibres and optoelectronic signalling

Immunity to electrical and magnetic fields

As photons have no electrical charge, they are unaffected by the electric fields found in high voltage environments or by electrical storms. They are also immune to endo- and exo-atmospheric pulse effects. In addition, high magnetic fields – caused by transformers and motors etc – have no effect on the fibre's optical transmission properties.

Low attenuation

An electrical signal transmitted over a coaxial cable will, dependent upon its frequency, have lost almost half its power after only a few hundred metres; whereas an optical signal transmitted over an optical fibre (quite frequently referred to as an **optical waveguide**) will, dependent upon the transmitted wavelength (see Chapter 2), still be perfectly acceptable after 20 kilometres.

A point to note when describing attenuation, is that it should *always* be made clear whether one is referring to an electrical decibel measurement or an optical measurement. This is because an electrical measurement of 1 dB is equivalent to an optical measurement of 2 dB. This consideration is extremely important when comparing cable attenuations (eg, 20 dB km^{-1} electrical cable = 10 dD km^{-1} optical cable). The reason for this is that optical power is directly proportional to electrical current, whereas electrical power is proportional to the square of the current. Summerised:

$$\text{dB [optical]} = 10 \log (P_{out}/P_{in}) = 10 \log (I_{out}/I_{in})$$

and;

$$\text{dB [electrical]} = 10 (P_{out}/P_{in}) = 20 \log (I_{out}/I_{in})$$

In addition, when specifying absolute optical power:

$$\text{dBm [optical]} = 10 \log (P_{out}/1\ \text{mW}).$$

Increased transmission bandwidth

At the moment, maximum transmission rates of about 1 Gbit s^{-1} are possible using conventional coaxial cable, whereas transmission rates in excess of 10 Gbit s^{-1}, are possible using optical waveguides.

Small physical size and weight

Another very important advantage associated with optical fibre cables is their comparatively small physical size and weight relative to coaxial cables. Whereas the diameter of a typical modern **coaxial cable** will be approximately 10 mm, a fibre rod is only 0·1 to 0·2 mm and even with a protective plastic coating this would not be more than 0·25 to 0·5 mm. In addition, the weight of a coaxial cable can vary anything from 350 to 1100 kg km^{-1} depending upon the type used, but a single fibre only weighs about 12 kg km^{-1} [8].

Previously the physical properties of copper cable have always severely restricted the size of cable that can be pulled through a duct and so these differences are very important when installation techniques are being considered.

Typically, a 1000-pair armoured copper cable would normally be manufactured in 400 m lengths and this would necessitate frequent jointing. The use of the far lighter fibre optic cable means that not only can longer lengths (eg, 2000 m [9]) be pulled through a duct with-

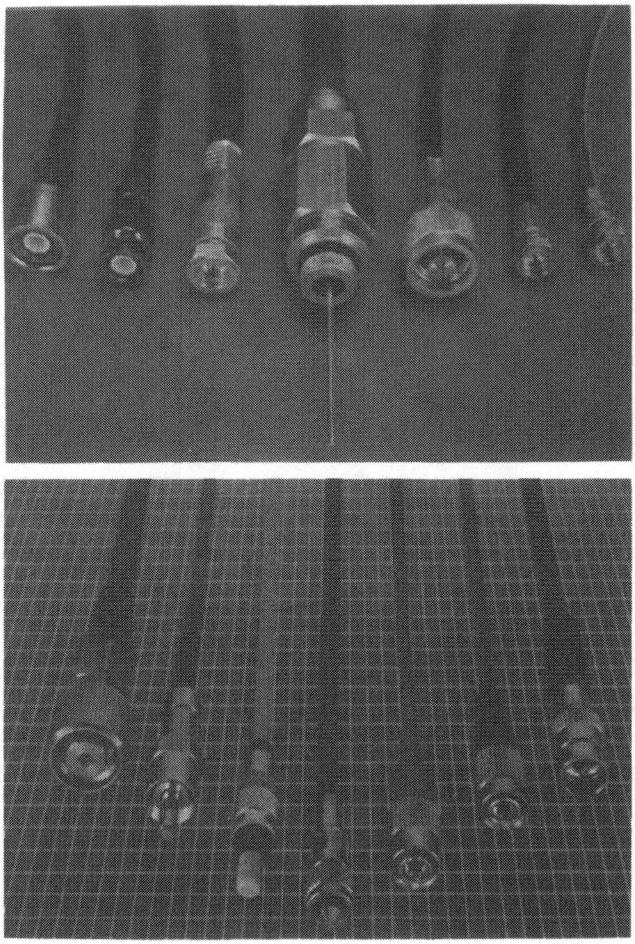

Plate 1.2 Comparison of coaxial (top) with optical fibre (bottom) cables and connectors (courtesy BICC)

out jointing, but transportation costs and problems associated with cable drums are also considerably reduced.

Increased flexibility

When designing telecommunications systems the increased flexibility of an optical fibre cable over a normal copper cable, together with its almost obsolescence-proof versatility, has to be considered.

In addition to being physically smaller, weighing less and possessing a relatively high **tensile strength** (approx-

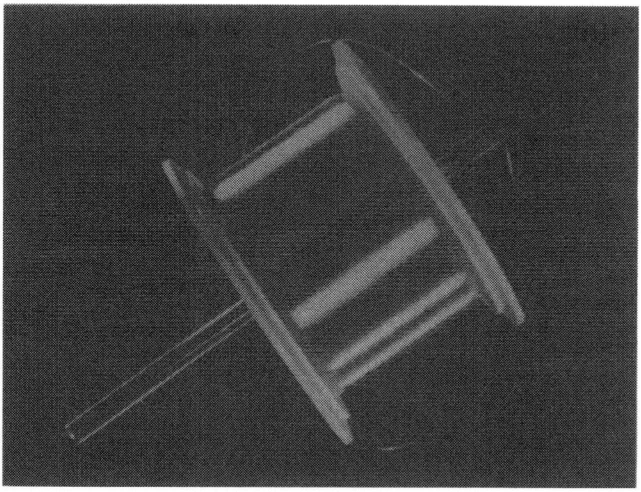

Plate 1.3
Optical fibre on a reel (courtesy Siemens AS Zentralstelle fuř Information)

imately equal to twice the kilometric cable weight and comparable, when freshly drawn, to that of any high tensile material, including steel [4]), optical fibres are far more mechanically flexible than normal copper coaxial cables. This is particularly advantageous with regard to transportation and in certain installations where space is at a premium (eg, aircraft). Optical fibres can also be incorporated into, or become part of, almost any existing network infrastructure with little or no difficulty.

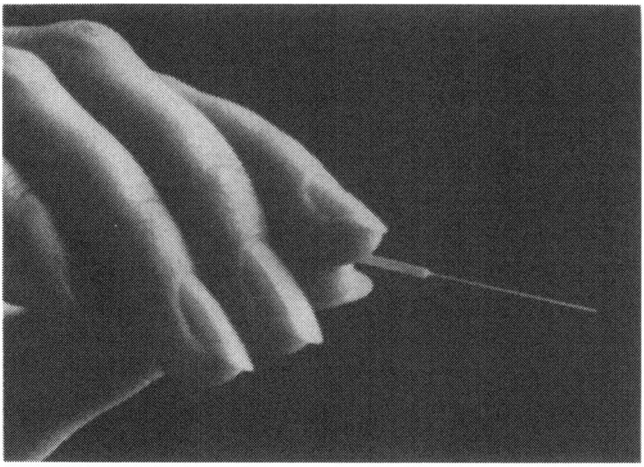

Plate 1.4
Single optical fibre cable (courtesy Siemens AS Zentralstelle fuř Information)

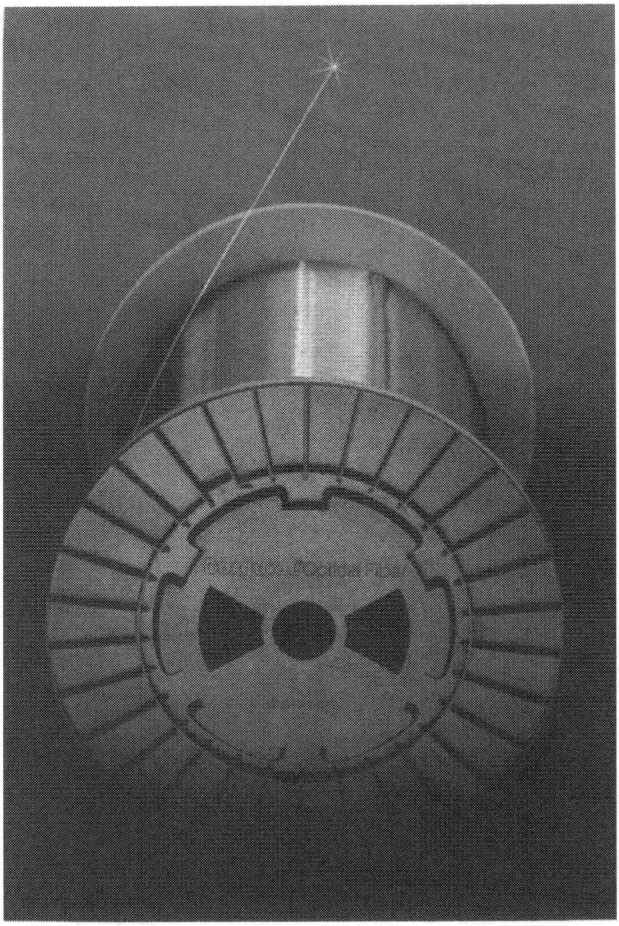

Plate 1.5
Reeled single-mode optical fibre (courtesy Corning Glass Works)

Electrical insulation

As an optical fibre is metal-free, there is no requirement to ground or earth the optical waveguide against, say, lightning strikes and, as no electrical signal is being transmitted, an optical waveguide is capable of bridging large differences in potential and could, for example, act as a control line for a high voltage switching station.

Electromagnetic interference and interception

As light waves carried inside a fibre do not set up any form of electromagnetic field, security of transmission is guaranteed and it is therefore virtually impossible to externally intercept a light signal.

The light wave is also unaffected by external electromagnetic interference and so the possibility of electromagnetic **crosstalk** (ie, the leakage of signal from one fibre to another) is eliminated. If optical leakage occurs, then it is retained in the plastic coating or cladding that surrounds the fibre, thus ensuring that optical crosstalk between fibres, cannot occur.

Another advantage of using fibre optics is that **tapping** a fibre can only be achieved by isolating the bare fibre from the cable and then completing an optical tap. Isolating the fibre and inserting a tap would naturally reduce the power output of the optical waveguide which in turn could be easily detected with the use of an **Optical Time Domain Reflectometer** (OTDR) or similar instrument (see Chapter 8).

Electrical protection

Because no electromagnetic field is present within a fibre rod, electrical protection (or **shielding**) is not required.

Analog and digital transmission

Both analog transmission (ie, the transmission of continuously varying signals either by direct modulation of the optical power or by using an electrical subcarrier to modulate the optical power) and digital transmission (ie, the transmission of a discrete signal, obtained by digitisation of an analog signal using binary light pulses in a periodic time pattern) are possible with the use of optical fibres. As the receiver requires a high **signal-to-noise ratio** for analog transmission and because **distortion** of the optoelectronic system is **non-linear**, digital transmission is a far more viable proposition. As a consequence of these limitations, analog transmission is normally restricted to lower bandwidths and smaller distances.

Receiver sensitivity

A further advantage of optoelectronic signalling is given by the inherent stabilisation of optical transmitters and the simulation in single-mode systems of modulation and demodulation processes (ie, heterodyning) whose effects are already well known from other electrical systems. Theoretical calculations have indicated that an increase of 10 to 15 dB in receiver sensitivity can be expected from the use of optoelectronic systems, compared with more conventional systems.

Disadvantages of optical fibres and optoelectronic signalling

Unfortunately there are also disadvantages associated with optical fibres and optoelectronic systems. These include:

Cost

Although the raw material used for producing optical waveguides (quartz sand) is inexpensive and available in unlimited quantities, the cost of obtaining a chemically-pure glass rod (typically 1 foreign atom in 10^9 silicon atoms) is particularly high because of the sophisticated manufacturing equipment and processes required.

Jointing and test procedures

Owing to the small size of the fibre compound to other cables, **micro-joining** procedures have had to be developed, together with their associated **test practices** (see Chapter 6 and 8 for more details). These are costly in manpower, training and equipment.

Short links

Even supposing optical fibre cable was inexpensive, it would still not be cost-effective to replace every small conventional connector (for example, between computers and peripherals) with an optical fibre link as the price of optoelectronic transducers is prohibitive. Currently, therefore, optical fibre links are used far more for long multi-transmission paths or high-density local area networks.

Fibre losses

Attenuation coefficients are a function of the optical wavelength, and so the amount of optical power available to the photodector at the end of a fibre length depends primarily upon the attenuation coefficient of the fibre.

Most of the energy losses in optical fibres occur because of **light scattering** due to microscopic imperfections in the glass, referred to as **Rayleigh scattering**. The degree of scattering increases as the light wavelength approaches the size of the imperfections, thereby giving greater attenuation at shorter wavelengths.

Impurity elements in the silica glass also produce high absorption at specific wavelengths, but as can be seen from the upper part in Figure 1.8 **windows** of low attenuation are found around 1300 nm and 1550 nm and it is these that are being exploited by modern infrared optical fibre systems. Above 1700 nm glass starts to absorb light energy due to the **molecular resonance** of silicon oxide. The lower part of the graph indicates the loss, limited by Rayleigh scattering.

Susanne Nagel, of AT&T Bell Laboratories, has suggested, in a recent paper, that the ageing effects of silica-based optical fibres occur mainly by radiation which causes cracks and increases attenuation. Nagel further points out that the cracks can be made much worse by hydrogen invasion.

Dispersion is another cause of energy loss and is brought about by light rays with different critical angles causing rays to have different path lengths. This can be overcome by using a fibre cladding (see Chapter 2).

Light propagation in a fibre is modal and similar to waveguide propagation [10]. In practice, however, cable

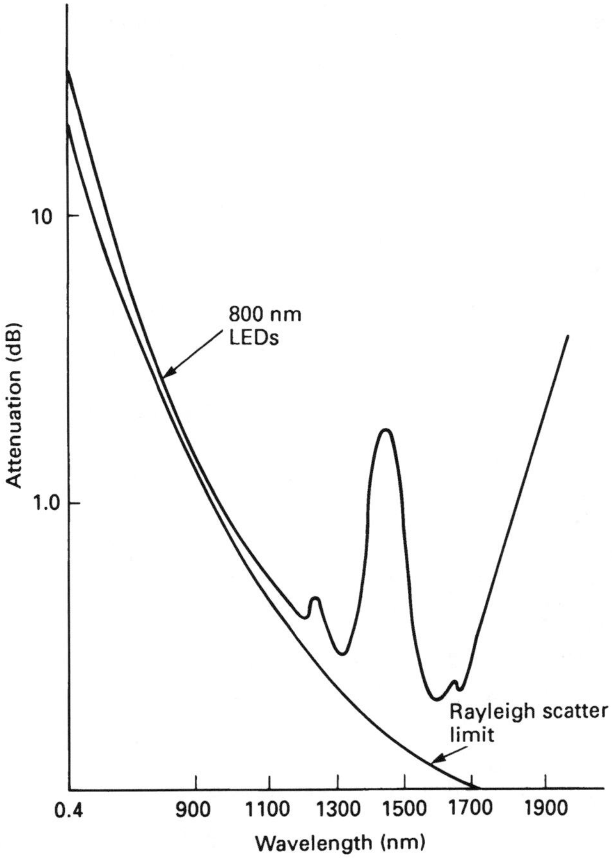

Figure 1.8 Attenuation of transmitted light in optical fibre, against wavelength. Windows of low attenuation are clearly seen when the light has wavelengths around 1300 nm and 1550 nm

diameters are very much larger than the optical wavelength, and therefore produce multimode propagation, with each mode having its own velocity. This leads to a dispersion or spread of energy in time as the light travels down the fibre thus limiting the maximum signalling speed over a given length of cable.

Because of fibre attenuation and material dispersion, optoelectronics at the moment is mainly concentrated in the **near infrared** or **short wavelength** (780 to 900 nm: typically 850 nm) and the **long wavelength** region (1200 nm to 1600 nm: typically 1300 nm and 1550 nm). Rayleigh scattering largely controls the **short**

wavelength losses, while absorption of the glass materials controls **long wavelength losses.**

Minimum loss is the region of about 1300 nm which is also the wavelength at which first-order material dispersion goes through zero (see Chapter 2).

Although designers would prefer to use the long wavelength region because their low fibre attenuation and comparatively low material dispersion permits higher bandwidths, the cost of doing so usually limits their choice.

Practical applications of optoelectronics

Telephone networks

Because of the low attenuation and high transmission bandwidth of an optical waveguide compared with copper lines normally used in telephone networks, more and more of the long-haul coaxial trunk links between telephone exchanges are being replaced by optical fibre links.

Urban broadband service networks

A normal coaxial cable television network using repeaters every 400 to 500 metres is only capable of transmitting 20 to 30 separate television channels within a total bandwidth of about 300 MHz. By using optical waveguides, not only can the bandwidth be increased, but the number of repeaters required is considerably reduced. This is particularly important in modern suburban communications where many information systems, for example **videotex, videoconferencing, videotelephony, switched broadband communication networks** (for extension of banking and shopping facilities) as well as **cable radio** and **television** can be supplied over a single fibre optic link.

The future of optoelectronics

In the not-too-distant future the technology associated

with optical line communication systems will have advanced enormously. This is particularly relevant with regard to system networks and the integration of components.

Switching networks

Optical transmission lines and **optical switches** (switching operations triggered directly by light pulses) are already at an advanced stage in the quest to seek replacements for electromagnetic and electronic switches.

Integration

Although integration is at the moment concentrating on **monomode optical waveguides** and the integration of a number of optical and electro optical elements on to a simple substrate via a thin film planer, the advancement of this technology will provide an enormous reduction in the cost of producing optoelectronic converters and repeaters etc.

As well as lower unit costs, however, there are many other advantages to integration; including higher packing density and increased reliability. But probably the greatest impact from integration has been with respect to **modulation** and **switching devices**, which are not only smaller, but now have a considerably reduced power consumption.

2
Theory

It was in 400 BC that Euclid was the first to notice that a pond always appeared shallower than it really was and he observed that light passing from one medium (such as air) into a denser medium (such as water or glass) always seemed to be bent (or refracted) by that denser medium.

In the second century Ptolemy, the Egyptian astronomer, measured the difference between the angle at which light was directed at a transparent medium (ie, the **angle of incidence)** and the corresponding angle that the light was refracted by that medium (ie, **the angle of refraction)**, shown in Figure 2.1.

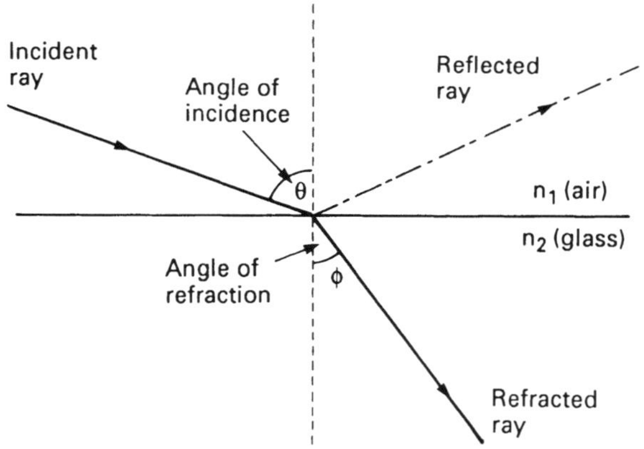

Figure 2.1
Showing how light is refracted when passing from one medium into a denser medium

Ptolemy later produced some tables of his findings and from these results he deduced that when a beam of light is directed at a transparent material such as glass, although some of the light is reflected back to the originating medium (for instance air) a high proportion of that light will pass into that material.

Ptolemy also termed **reflectance** as being the ratio of reflected power to incident power.

Refractive index

The optical density of a material is referred to as its **refractive index** and is a direct proportion between the velocity of light in a vacuum to the velocity of light in the material:

$$\text{refractive index} = \frac{\text{velocity of light in a vacuum}}{\text{velocity of light in a medium}} = n$$

The refractive index of glass varies between the limits of pure **crown glass** and pure **flint glass** with a typical value of 1.5.

Snell's Law

In 1621 Willebrord Snell (1561–1626) put forward a paper which, although never actually published, postulated that the **refracted ray** will always lie in the same planes as the **incident ray** and that the sine of the **angle of refraction** (ϕ) is dependent upon the sine of the **angle of incidence** (θ) such that:

$$\frac{\sin\theta}{\sin\phi} = \frac{n_2}{n_1}$$

where n_1 and n_2 are the indices of refraction of the two materials through which the light is passing.

This law of refraction is generally known as **Snell's Law** although it was not until some years later that a Frenchman called René Descartes was the first to really make full use of the ratio between the sines of the two angles. Because of this it is not uncommon, especially in France, to hear Snell's Law referred to as **Descartes Law.**

If, as in Figure 2.2, a ray of light is now directed at a

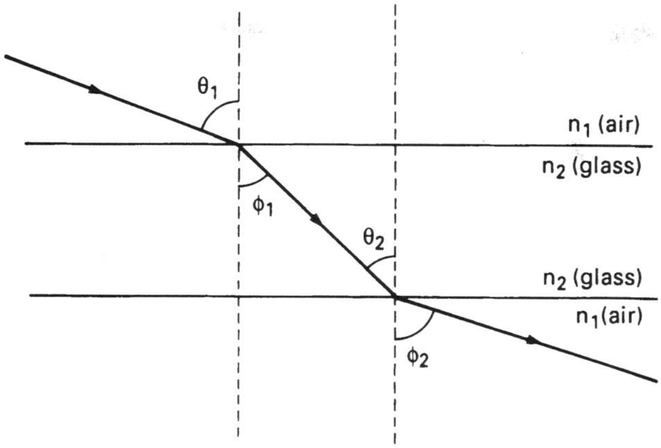

Figure 2.2
Showing the path of a ray of light through a block of glass with parallel sides

block of glass, the angles ϕ_1 and θ_2 must be equal because the sides of the glass are parallel. So:

$$n_2 \sin \phi_1 = n_2 \sin \theta_2$$

and, therefore:

$$n_1 \sin \phi_2 = n_1 \sin \theta_1$$

Thus the angles θ_1 and ϕ_2 must always be equal.

If, however, the light source is now contained *inside* a glass tube (Figure 2.3) then, by Snell's law, the light leaving the source strikes the surface of the glass at an angle of θ and leaves an angle of ϕ, equivalent to:

$$n_2 \sin \theta = n_1 \sin \phi$$

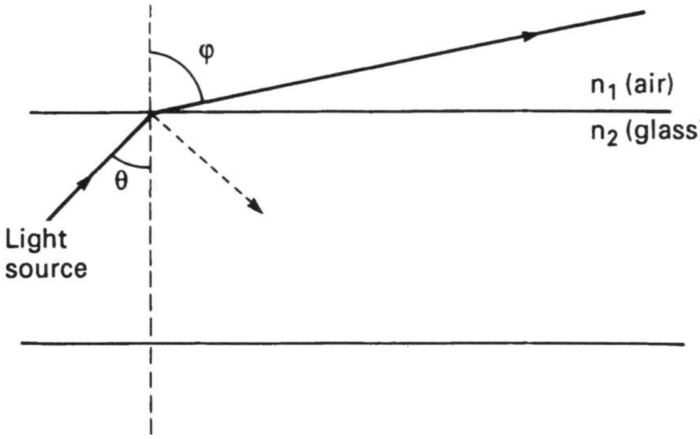

Figure 2.3
Situation with a light source within the glass tube

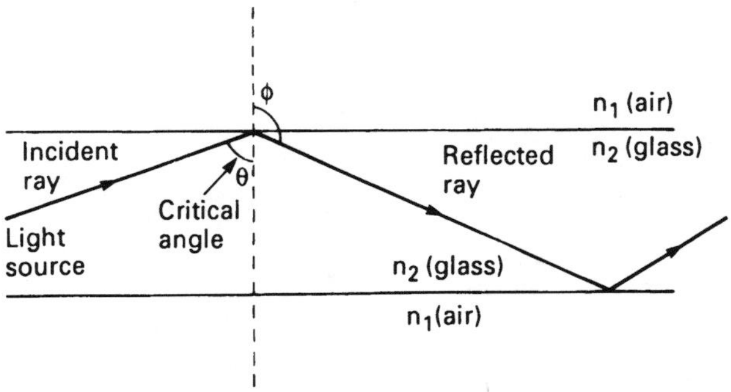

Figure 2.4
Showing total internal reflection of a light ray within a glass tube

As n_2 is greater than n_1 then ϕ must be greater than θ and, if ϕ is greater than 90 degrees, there will be **total internal reflection** (ie, *no* refraction) and all light will remain inside the glass tube (Figure 2.4). Furthermore, as the sides of the glass tube are parallel, rays of light striking the surface will be internally reflected at an angle equal to the angle of incidence. Reflected rays will be reflected at exactly the same angle as they are incident on the opposite side. This will continue until they reach the far end of the tube.

The glass tube therefore acts as a **guide** for light and, as the light is trapped inside, bends in the glass tube are relatively insignificant. Scratches or surface contamination such as grease, however, alter the effective angle of incidence at that particular interface. This is because grease has a refractive index close to that of glass and hence to the approaching ray this appears as an extension of the glass surface.

The angle of incidence that causes an angle of refrac-

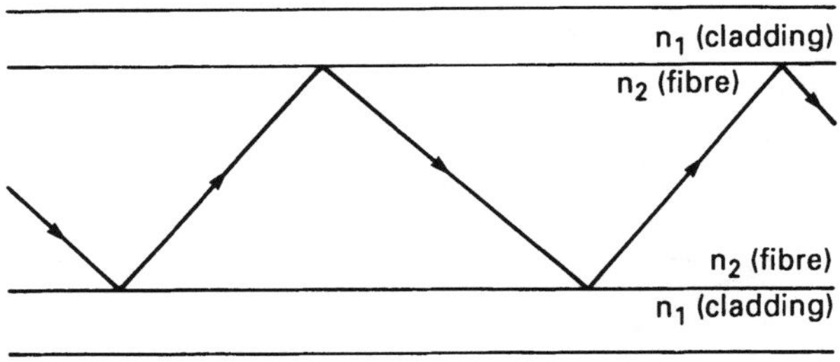

Figure 2.5
Critical angle, for a tube encased in cladding with only a slightly lower refractive index the critical angle, is reduced, resulting in lower dispersion loss

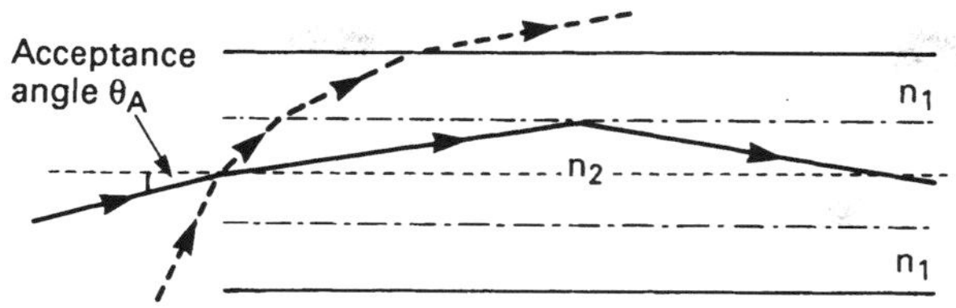

Figure 2.6 Showing the acceptance angle of light incident on the end face of an optical fibre

tion greater than 90 degrees is called the **critical angle** and, for glass (with a refractive index of about 1.5) in air, the critical angle is approximately 42 degrees. By encasing the glass tubes inside a **cladding** whose refractive index is only slightly lower than the refractive index of the glass (Figure 2.5), the critical angle – and hence any dispersion loss – is reduced.

This is the basic principle behind the use of optical fibres and optoelectronics, and is covered by the following rules:

- when light propagates, in a glass of refractive index n, it moves by a factor of n more slowly than in free space
- if light emerging from a glass with a high refractive index is incident on a material with a lower refractive index, it changes direction provided it strikes the interface at a steep angle. If the angle of incidence is shallower than the critical angle, total internal reflection occurs
- when a light beam is passed from air to glass, because of the different refractive indices at the point of penetration, the light is propagated from within the glass core at various angles. A **variable refraction** is therefore present at the end face. This is sometimes referred to as the **aperture** [11][12].

Acceptance angle

The **acceptance angle** is a function of the refractive indices of the core and cladding materials and is defined as: the half angle of vertex of a cone within which rays are incident on the end face of the fibre and will be captured and propagated by that fibre (Figure 2.6).

Plate 2.1 Multimode fibre propagation, showing the numerous propagations as light rays emitted from a fibre (courtesy of Corning Glass Works)

Numerical aperture

The sine of the acceptance angle is called the **numerical aperture** of the fibre given the symbol A_n, and is more commonly quoted than the acceptance angle itself.

Modes

Many different types of fibres are manufactured today (see Chapter 3) but there are only two modes of propagation – **single-mode** (often called **monomode**) and **multimode.**

As the name suggests, in a single-mode fibre, only one mode, the **fundamental** or monomode, can propagate. In multimode fibres, however, several hundred modes differing in field pattern and propagation velocity are possible. The upper limit to the number of modes is determined by the core diameter and numerical aperture of the waveguide. For example, a graded index fibre with a 50 µm core typically allows about 500 separate core modes each having different propagation and attenuation characteristics.

Optical power and power density

A uniformly diffused light emitting surface is called a

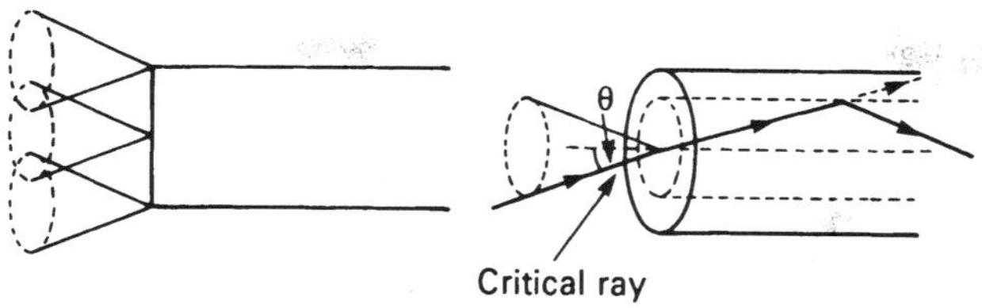

Figure 2.7
Any point on a fibre's end surface accepts incident power within its acceptance cone

lamberton source and when a surface is radiating power, the power density at any point on that radiating surface can be expressed in terms of the power per unit area (normal unit: Wcm^{-2}). This is called the **irradiance** of the source and its distribution across the surface gives the **near field radiation pattern.**

The power radiated in a given direction is referred to as the **radiant intensity** of the source and its distribution forms the **far field radiation pattern.**

Power density decreases with source distance until it eventually becomes almost insignificant. The density of this radiant power is expressed as the **radiance** of the source and the unit that describes the optoelectronic radiation characteristics is called the **steradian.**

Optical fibre input power

If a large radiant source such as an LED is in close proximity to the surface area of a fibre end surface (see Figure 2.7), each point on that fibre end surface accepts incident power within its **acceptance cone,** and the total of all such incremental powers gives the total power input.

Dispersion losses

The dispersion or spreading of pulses as they travel through a fibre results from two factors: the broad range of wavelengths emitted by the light source and the wavelength dispersion inherent in the fibre.

Material dispersion

The amount of material dispersion depends upon the chemical composition of the glass. As the refractive index of a glass prism varies with wavelength, different wavelengths will travel at different velocities along the fibre and arrive at different times. This causes an energy loss which is called **material dispersion**.

Waveguide dispersion

As frequency is a function of wavelength, the **group velocity** of the energy in a fibre varies with frequency. This produces an additional loss which is called **waveguide dispersion** and arises from the distribution of light between the case and the cladding.

Chromatic dispersion

The combination of material dispersion and waveguide dispersion is called **chromatic dispersion** (often called **wavelength dispersion**) and losses can be minimised by reducing the spectral width of the transmitter and through choice of the correct wavelength.

The light emitted by a typical red LED covers a waveband of approximately 40 nm from (780 nm to 820 nm). In contrast, the most recent laser diodes have a spectral width of less than 1 nm. Choice of transmitter is thus one of the main methods of reducing spectral width.

Choosing the correct wavelength is, however, equally important. Figure 2.8 shows a graph of effective refractive index against wavelength, illustrating the effects of material, chromatic and waveguide dispersion. Material dispersion and waveguide dispersion effects vary in

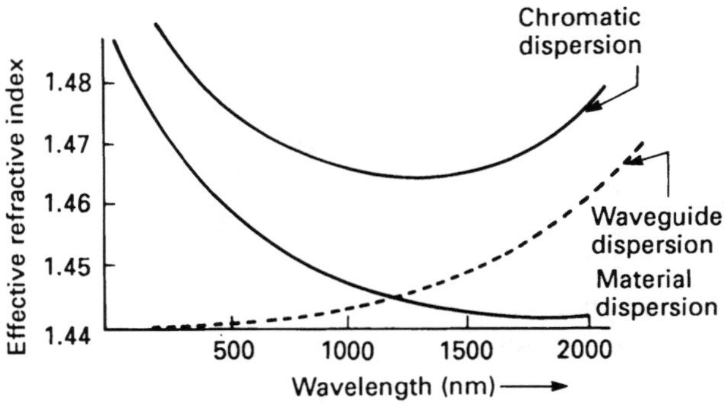

Figure 2.8
Graph of an optical fibre's effective refractive index against wavelength, showing effects of chromatic dispersion, waveguide dispersion and material dispersion

opposite senses as the wavelength is increased, but there is an optimum wavelength around 1300 nm at which the two effects almost cancel each other out and chromatic dispersion is at a minimum. Attenuation is therefore also at a minimum and this makes 1300 nm a highly attractive operating wavelength and a point at which there is a possibility of increasing propagation path lengths to hundreds of kilometres.

Reducing chromatic dispersion

By increasing the waveguide dispersion so as to cancel out material dispersion, it is possible to shift the zero-dispersion wavelength to a 1550 nm loss minimum.

The first, commercial, dispersion-shifted fibre was introduced in 1986 by Corning. This has a segmented core whereby a high-index inner region is surrounded by a lower-index segment, a segment with a higher refractive index and finally a low-index cladding.

There are a number of other methods of achieving the same result and these include reducing the core diameter or making the core's refractive index have a triangular, trapezoidal or trenched-annular radial distribution.

Although production runs of segmented-core dispersion-shifted fibres have shown average losses of 0.21 dB km^{-1} at 1550 nm, current costs for dispersion-shifted fibre are higher than standard step-index fibre, but this difference is expected to narrow as production volumes increase [13].

Optical transitions in semiconductors

The basic construction of **semiconductor devices** are such they have two **electron energy bands (valance** and **conduction)** separated from each other by a **band gap**, E. The interaction between photons (ie, light particles) and valence/conduction band electrons, is the basic principle behind all optoelectronic semiconductors.

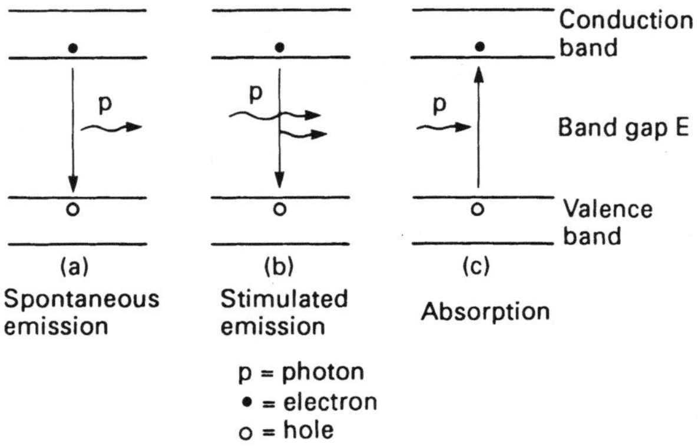

Figure 2.9 Semiconductor transitional mechanisms

Semiconductor optical transitional mechanisms

There are three semiconductor optical transitional mechanisms, these are spontaneous, stimulated emission and absorption, illustrated in Figure 2.9.

When a conduction band has too many electrons, the electrons are passed to the free states (holes) in the valence band. This causes a radiative recombination of excess charge carriers and a photon (p) is spontaneously emitted whose energy is proportional to its frequency and wavelength and is thus dependent upon the size of the band gap E (Figure 2.9a). This is called **spontaneous emission** or **luminescence.**

Each band occupies a range of energy values which means that the wavelength can vary from photon to photon and thus provide a bandwidth to the total emission.

For germanium and silicon the band gap is such that emission is in the far infra-red or heat band, but some compounds of band III-V elements can produce emissions of visible or near infra-red wavelengths.

Because of photons stimulating excess charge carriers to radiative emission, additional photons are emitted. This is called **stimulated emission** (Figure 2.9b).

The third optical mechanism that can occur is when an electron is being taken from the valence to the higher energy state or conduction band and the photon is overcome by the electron. This is referred to as **absorption** (Figure 2.9c).

Although absorption, spontaneous emission and stimulated emission always occur together, one type of emission will be dominant.

Forward biasing the semiconductor p-n junction will, similar to a transistor, cause charge carriers to move from one side of the junction to another where, as minority carriers, they will be recombined with the majority carriers. This phenomena of emitting light after forward biasing is called **injection luminance**. The devices that provide this particular type of interaction are **light emitting diodes**, **laser diodes** and in the reverse mode, **photodiodes**.

Usable bandwidth of optoelectronic transmission systems

To transfer a digital waveform over a normal voice telephone system, the waveform must first be converted into signals that fall within the frequency of the speech channel ie, 300 to 3400 Hz. This can be carried out by means of a **modem** (modulator/demodulator) which is connected between the terminal and the telephone line. Because of the limitations of a normal telephone speech channel (eg, **line distortion**, noise and the requirement to prevent **interference** to adjacent channels), the transmission of data at bit rates only up to 2.4 Kbit s^{-1} is normal. Of course, if high quality private wires or 4-wire circuits are used, rates of up to 9.6 Kbit s^{-1} and in exceptional cases 19·2 Kbit s^{-1} are feasible; depending upon the **bit error rate** (ie, the number of false bits relative to the total number of bits in a digital stream).

Bit error rate is recognised as the quality measurement of a transmission system and typical requirements are 1 in 10^9 but could be as high as 1 in 10^{13}.

In digital communications, one speech channel utilises a 64 Kbit s^{-1} high speed data channel: to perfectly **pulse code modulate** a speech channel requires 64 Kbit s^{-1} (**Delta modulation**, incidentally, only requires 16 Kbit s^{-1}

and **digital analysis** only 2.4 Kbit s^{-1}.) All of these limits are covered by recommendation *CCITT Signalling System No 7* [14], which stipulates that digital signalling shall be via a 2 Mbit s^{-1} stream (actually 2048 Kbits), split into 32×64 Kbit s^{-1} traffic channels.

Depending upon the acceptable amount of voice degradation, this can be reduced to about 48 Kbit s^{-1}, which also enables 16 Kbit s^{-1} of data to be transmitted at the same time.

Research establishments are now investigating the possibility of utilising 144 Kbit s^{-1} channels to allow a number of speech and data channels to be transmitted simultaneously. This will provide:

- 2×64 Kbit s^{-1} speech channels.
- 1×8 Kbit s^{-1} data channel
- 1×8 Kbit s^{-1} high speed data channel

Currently a range of all digital services are being offered by PTTs worldwide. These are available at 64 Kbit s^{-1} and 2.04 Mbit s^{-1} bandwidths (British Telecom's services are called ***KiloStream*** and ***MegaStream***), but availability of higher speeds is only a matter of time.

Cut-off frequency of an optical fibre

The cut-off frequency of an optical fibre is determined not only by the fibre itself (mainly modal dispersion in the case of multimode fibres, and waveguide dispersion in the case of single-mode fibres) but also by the amount of material dispersion caused by the spectral width of the transmitter [15] (see earlier).

Fibre bandwidth

The frequency at which signal attenuation increases by 3 dB is called the fibre bandwidth. As the bandwidth of an optical waveguide is virtually reciprocal to its length, the bandwidth length product is frequently quoted as a **fibre quality characteristic**. Figure 2.10 shows frequency responses of 1 km lengths of a number of different types of

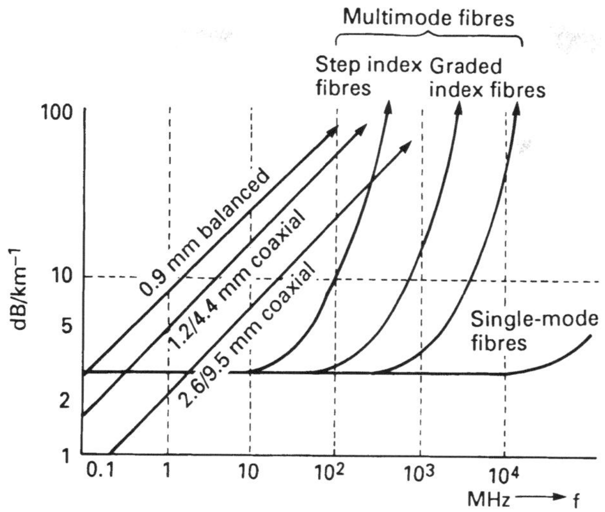

Figure 2.10 Typical frequency responses of optical fibres compared with copper cables

fibres compared with 1 km lengths of balanced copper and coaxial cables. It should be noted, that the fairly constant response of an optical fibre obviates the need for **equalisation** which is normally required for coaxial systems.

Owing to material dispersion, for fibres greater than 1 km in length, the cut-off frequency is also dependent upon the actual material used and the number of splices.

Modulation and demodulation of the sub-carrier

The principle of signalling over a copper cable relies on an oscillator being modulated by the information signal in frequency, phase or amplitude. The product is then demodulated by a detector at the distant end.

The transmission methods used for optoelectronics mainly depend upon whether the signal is analog or digital, but the principles are similar.

Frequency modulation

If the optoelectronic transmission path bandwidth is much greater than the signal being transmitted, a suitable frequency modulated electrical sub-carrier can be introduced.

By modulating a sub-carrier it is possible to create an analog transmission channel which could be suitable for transmitting video and frequency division multiplex (FDM) signals as well as digital signals.

Several analog signals (eg, video channels) can be combined to form an FDM signal which can then be used to frequency modulate the electrical sub-carrier. This method would however reduce the dynamic range of the channels and a far better method would be to frequency modulate a number of carriers with video signals and then to multiplex these on a frequency division basis.

Intensity (power) modulation

Because the spectral width of light emitting and (to a lesser extent) laser diodes is so high and the frequency stability so low, frequency and phase modulation of optical transmitters is almost impossible. Another disadvantage is associated with the fibre itself (owing to the multipath propagation properties of step index and graded index multimode fibres, the signal is greatly distorted). Nowadays, therefore, the only practical type of modulation that can be used is **intensity (power) modulation** whose main advantage is that the optical source can be directly modulated by the drive current and the photodiode will convert the received optical power directly into an electrical photocurrent. There are two main types of intensity modulation: direct intensity modulation and premodulation with subsequent intensity modulation.

Direct intensity modulation is a very simple process which relies upon the optical signal and the electrical signal being proportional. Provided that the transmitter and the receiver characteristics are exactly the same, a high linearity can be realised; but as this can only normally be achieved in small sections, its use is restricted mainly to analog transmission although it is sometimes used for transmitting signals over short distances.

With **premodulation and subsequent intensity modulation** the advantages are that there is an improved signal-to-noise ratio and the non-linearity of the optoelectronic transducer no longer affects the signal. The design is again simple and this method is particularly suitable for transmitting video signals over short or medium distances.

Usable bandwidth of an intensity (power) undulated optoelectronic transmission path is determined by:

- the **modulation bandwidth** of the optical transmitter
- the **dispersion effects** in the fibre (eg, modal, material and waveguide)
- the modulation bandwidth of the optical receiver [16].

Signal-to-noise ratio

The signal-to-noise ratio at the output of an optoelectronic transmission systems depends upon:

Optical receiver thermal noise

This is noise due to the **load resistor** and the first amplifier. This noise can be minimised by utilising a device such as a **transimpedence amplifier** to reduce the effective time constant and diode capacitance, which limits the receiver frequency response.

Quantum noise (of light)

This is mainly due to the noise caused by **avalanche diodes.**

Dark current noise

Even when the photodiode is at rest a reverse current flows and this produces a noise component. Although this effect can be reduced by using an avalanche photodiode, avalanche diodes themselves, produce noise and so there is a limitation to their use. Dark current noise is normally reduced in optical receivers by using a low pass filter.

Laser noise

The actual structure of the laser can effect the stability of the transmitter's output power.

Mode partition noise

This is caused when individual **laser modes** arrive at the receiver separated in time and produce an increase in the noise.

Modal noise (in multimode fibres)

When a graded index multimode fibre (see Chapter 3) is used together with a narrow spectral-width laser diode, the interference pattern produced is very unstable owing to fibre effects and fluctuations in the optical wavelength. This is particularly noticeable at junctions and splices.

It has been ascertained [15], that at low powers **receiver noise** (ie, thermal and quantum light) predominates and the signal-to-noise ratio will improve with optical power until **transmitter noise** (ie, laser noise, laser modal noise and fibre modal noise) becomes dominant and the signal-to-noise ratio then remains constant, as shown in Figure 2.11.

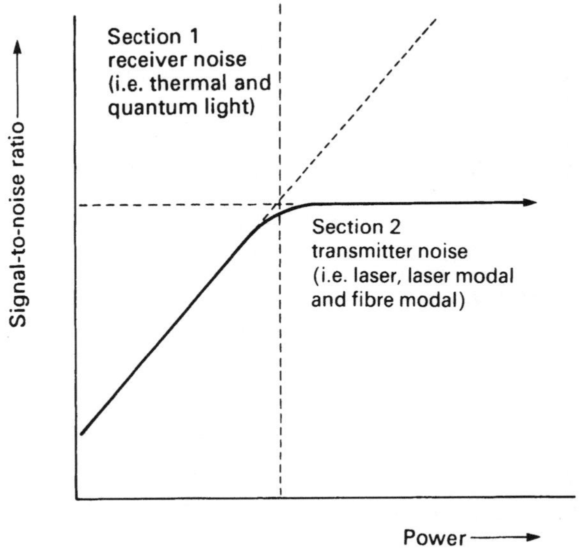

Figure 2.11
Graph of signal-to-noise ratio against power at the output of an optical receiver

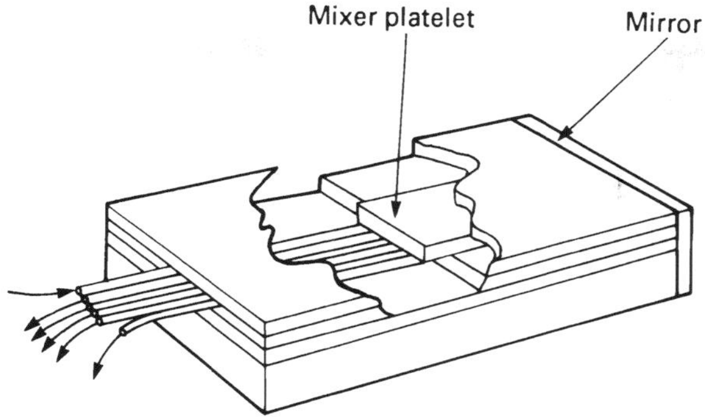

Figure 2.12
End fire mixer with platelet and mirror

Signal-to-noise ratios of around 40 dB can be expected with **gain-guided laser diodes** (see Chapter 4).

Types of mixers

End fire mixers

A thin platelet of **quartz glass**, embedded in a layer of low refractive index adhesive, whose thickness is equal to the core diameter of the fibre sections and whose width is the same as the total outer diameters of the fibres, ensures that incoming light is equally divided to outgoing fibre ports and therefore acts as an effective waveguide (Figure 2.12). A mirror reflects the light.

Transmissive mixer

This is similar to the end fire mixer, except that the incoming and outgoing fibres are arranged opposite each other (Figure 2.13).

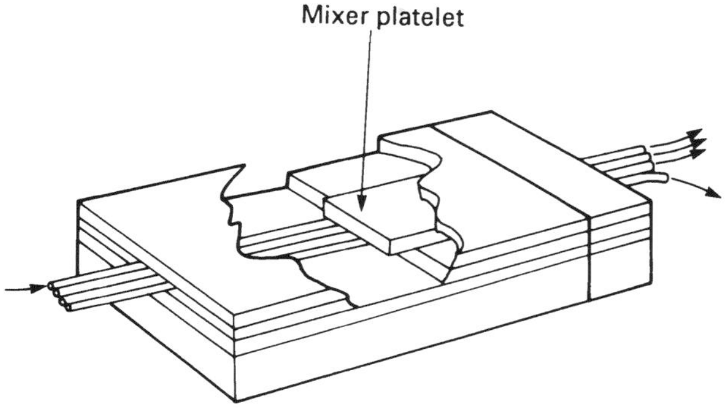

Figure 2.13
Transmissive mixer, with mixer platelet

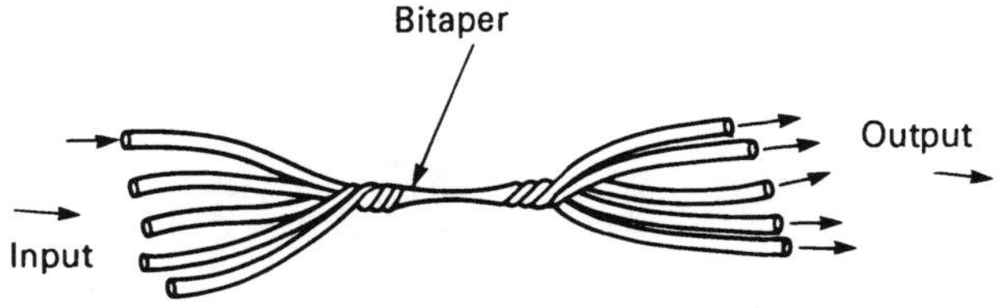

Figure 2.14
Biconical tapered transmissive mixer

Biconical tapered transmissive mixer
In the biconical tapered transmissive mixer, the power from any or all input fibres is spread among all of the output ports (Figure 2.14).

Types of multiplexing
Owing to the dispersion characteristics of optical waveguides and because of financial restraints, quite often it is better to use a high quality single-mode fibre (see Chapter 3) with a transmission bandwidth of 565 Mbit s^{-1}, as opposed to multiplexing four 140 Mbit s^{-1} signals together.

Currently there are four main methods of multiplexing several electrical signals together via an optical fibre link.

Fibre multiplexing
This is where a separate fibre is used for each signal.

Electrical multiplexing
Electrical multiplexing is where as many as four input signals are first combined electrically and the resultant signal used to drive an optoelectronic transducer.

Although electrical multiplexing is much more economic and considerably less complex than wavelength division multiplexing (see later) for normal telecommunications links, the use of wavelength division multiplex for subscriber line networks is advisable due to the greater flexibility provided for the gradual expansion of installed networks.

For instance, in an initial stage, narrow band sources such as telephone and data transmission can be transmit-

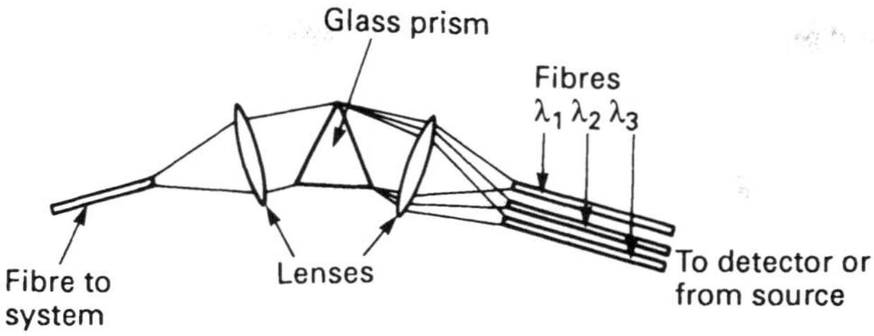

Figure 2.15 Wavelength division multiplexing using a prism

ted over a single fibre to the subscriber using bidirectional wavelength division multiplex in the range 800–900 nm. If at a later date visual communication is required, then signals can be transmitted in the 1300 nm region over the same fibre by the use of additional optoelectronic transducers.

Wavelength division multiplexing

Wavelength division multiplexing is the the simultaneous transmission of several signals in an optical waveguide at different wavelengths.

Several techniques of achieving this exist, such as using a prism as a wavelength dispersion element (Figure 2.15) or using a multilayer dielectric filter as a wavelength selective beam splitter (Figure 2.16). The devices which perform wavelength division multiplexing and demultiplexing are commonly called **couplers.**

In the example system of Figure 2.17, three ports on each coupler are used – and the coupler acts as either a wavelength division multiplexer or a demultiplexer if used in the reverse mode. This sort of coupler is called a **muldex coupler**: a contraction of the words multiplexer and demultiplexer.

This technology enables signals that are independent

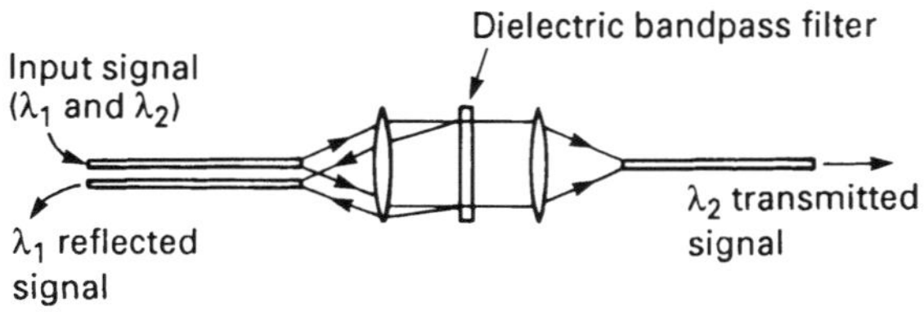

Figure 2.16 Wavelength division multiplexing using a multilayer dielectric filter

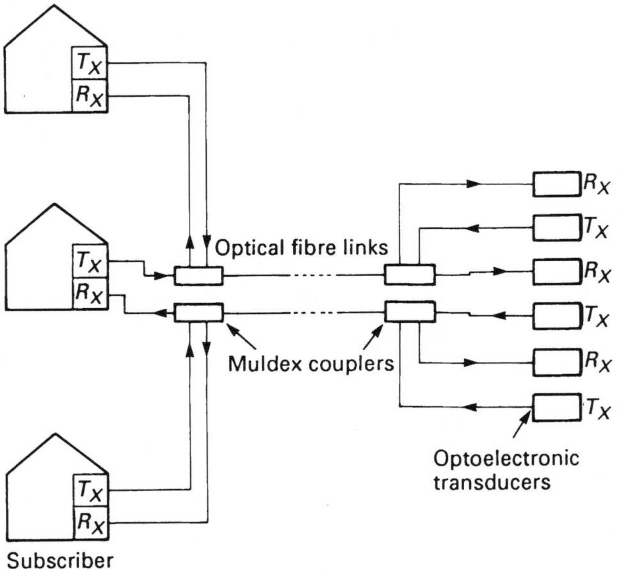

Figure 2.17 An example of a wavelength division multiplexed optical fibre transmission system, with bi-directional muldex couplers

of each other, at different carrier wavelengths and close channel spacing to be modulated in a different manner (ie, digital or analog) and transmitted over the same fibre. **Insertion losses**, however, are increased; which shortens the span length to about 5 km and this type of multiplexer is used mainly in local **TV subscriber networks.**

Time division multiplexing

In a time division multiplexed system, a number of sampled messages are used to modulate a pulsed carrier at different time intervals. This ensures that only one carrier is actually transmitted at any one instance.

3
Fibres and cables

Two types of fibre are used for optical transmission (step-index and graded-index) and two modes (single-mode and multimode).

Single-mode fibre

Single-mode fibre (also known as **fundamental** or **mono-mode** fibre) will permit only one mode to propagate and, as such, cannot suffer **mode delay differences**. Single-mode fibres are capable of wide bandwidths, for example, up to 40 GHz [17] and are, therefore, ideally suited for long-haul and high capacity circuits.

Figure 3.1 shows single-mode fibre, together with its refractive index profile and cross-section. Although the outer cladding of an optical fibre cable is at least ten times the thickness of the core radius (so as to prevent micro-bending losses), the actual fibre core is only a few micro-

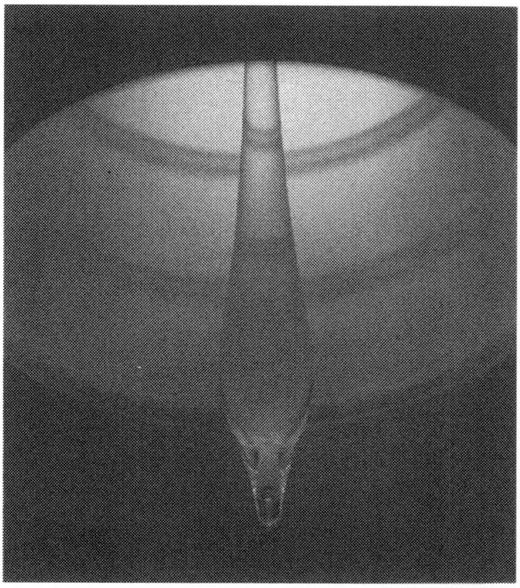

Plate 3.1
A glass 'gob' falls free of the heated end of a silica blank, thus beginning the fibre drawing process – see Forming the glass rod and Fibre drawing (courtesy of Corning Glass Works)

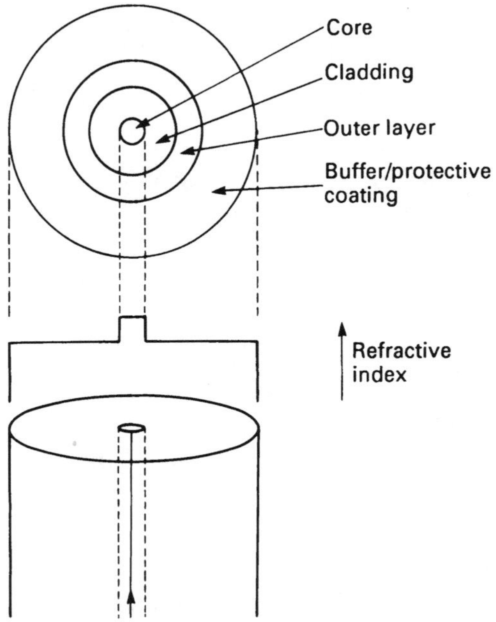

Figure 3.1 Single-mode optical fibre, together with its refractive index profile and cross-section

metres in diameter (typically, 2 to 8 μm) and there is a refractive index difference of less than 0.3 per cent.

The unfortunate disadvantage of using a single-mode fibre is the difficulty experienced in trying to couple the extremely small cross-sectional area of the fibre core to an optoelectronic detector or when trying to splice it to another length of fibre.

If the fibre is subjected to mechanical stress, local discontinuities can be introduced. Curvatures of the fibre involving axial displacements of a few micrometres and spatial wavelengths of a few millimetres result in unwanted light radiance and extra attenuation losses. These are referred to as **microbending losses**, or in the case of macroscopic axial deviation of the fibre from a straight line, **macrobending losses.**

To overcome these problems, high-intensity laser diodes are frequently used.

Another disadvantage of single-mode fibres is that as the refractive index of glass decreases with optical wave-

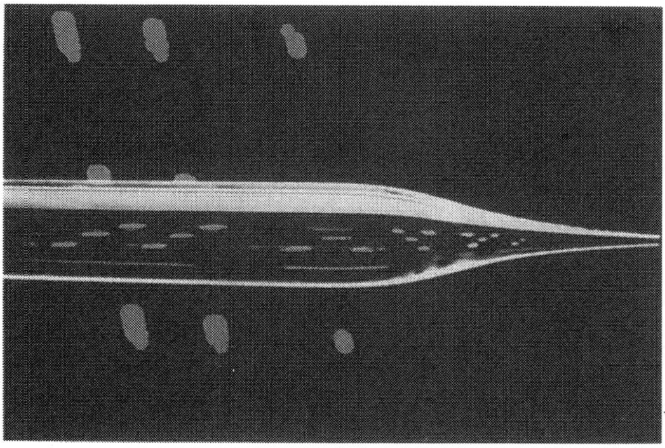

Plate 3.2
The consolidated glass blank – see Forming the glass rod and Fibre drawing (courtesy of Corning Glass Works)

length, the light velocity will also be wavelength dependent and because the light from an optical transmitter will have a definite **spectral width** (eg, LEDs: 40 nm, multimode lasers: 3 nm, monomode lasers: 0.01 nm), delay differences will occur, even in a single-mode fibre. There is therefore an advantage to using a (limited) multimode fibre.

Multimode fibre

In comparison to the single-mode fibre, multimode fibre has a relatively large core (typically 50–60 μm) and a high numerical aperture. As the name implies, multimode fibres are capable of propagating more than one mode at a time and they are ideally suited for high bandwidth (ie, a few GHz), medium-haul applications. The outer cladding is normally ten wavelengths thick and approximately twice the size of the core diameter. There are two types of multimode fibre.

Step-index fibre

As stated in Chapter 2, a small glass fibre in air has a large critical angle of approximately 42 degrees. This enables propagation along paths that are much longer than the axial route but attenuation and dispersion losses are subsequently increased.

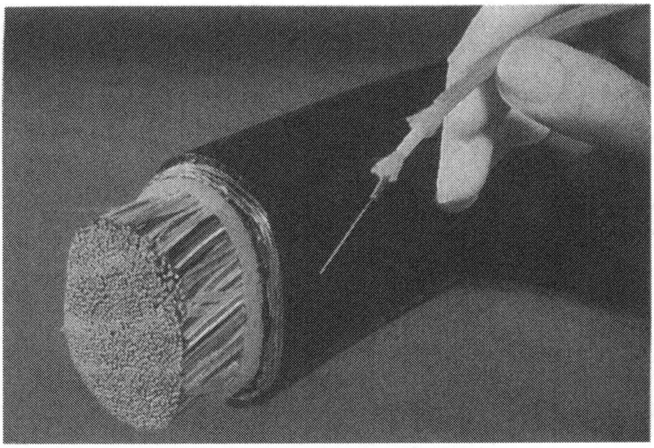

Plate 3.3
A single-fibre cable can carry more than 12,000 simultaneous telephone conversations. The large twisted-pair copper cable has a capacity of only 256 (courtesy of Corning Glass Works)

By making the fibre core extremely narrow (*CCITT Recommendation No G651* states that the core diameter should be 50 µm and the cladding 150 µm) and enclosing it in a cladding material whose refractive index is only slightly lower than the fibre (but nevertheless capable of ensuring that the refractive index changes directly at the core cladding interface) will not only reduce the critical angle but also reduce the losses.

A fibre optic constructed in this manner is referred to as a step-index fibre and they are ideally suited for short-haul and low-cost installations with typical bandwidths up to about 20 MHz [17]. Figure 3.2 shows a step-index multimode fibre, together with its refractive index profile and cross-section.

In a simple step-index fibre the refractive index of the cladding is typically 10 per cent lower than the refractive index of the core and, provided that the light coupled into the fibre is of a low acceptance angle, total reflection of light rays will occur. It should be noted however that although the light velocity will be constant within the core, a ray with a high angle will have smaller velocity component than those with a smaller reflective angle. An angle of 12 degrees is usually considered as a typical value of acceptance angle.

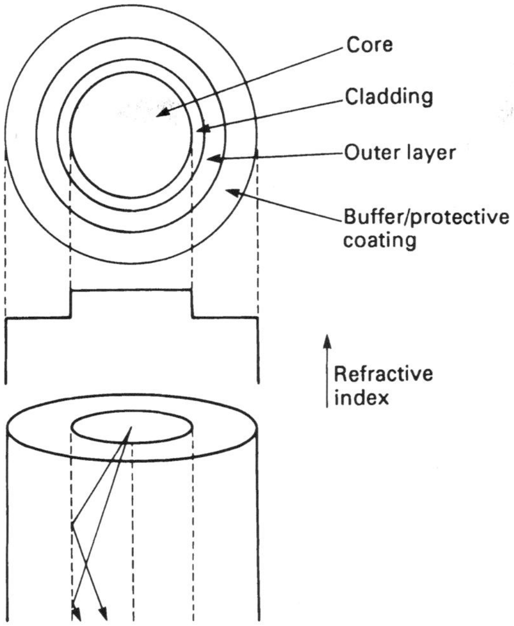

Figure 3.2
Step-index multimode fibre, together with its refractive index profile and cross-section

The main disadvantage of step-index fibres is that the different optical lengths caused by the various angles at which light is propagated relative to the core, causes the transmission bandwidth to be relatively small. These effects can be avoided by using graded-index fibre.

Graded-index fibre

Although previously only used for long-haul (trunk) networks, graded-index fibres are nowadays frequently being employed in local networks at the 850 and 1300 nm wavelengths.

In a graded index fibre, the core's refractive index is made to decrease *parabolically* with radial distance and because of the variations in refractive index, the trajectory of the light is *sinusoidal* as opposed to zigzag and the light rays follow curved paths. Figure 3.3 shows graded-index fibre, together with its refractive index profile and cross-section. Figure 3.4 shows the light trajectory in detail.

Figure 3.3
Graded-index multimode fibre, together with refractive index profile and cross-section

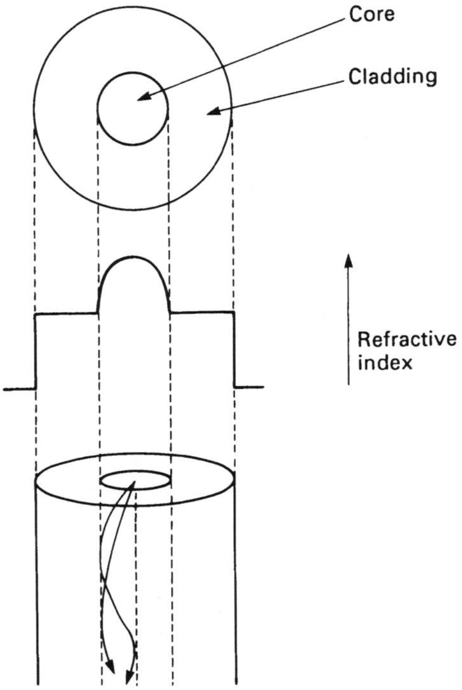

The reason for this is that the core is, in effect, made up of a central membrane or **fibre axis**, surrounded by a succession of infinitely thin shells each consisting of a different, but constant, refractive index. From Figure 3.4 it can be seen that light rays running close to the fibre axis, although having a shorter path length, will have a lower *velocity* because they pass through a region with a high refractive index.

Owing to the reduced refractive index at the core edges the rays on more deviant paths will travel much faster

Figure 3.4
Light trajectories in a graded-index multimode fibre

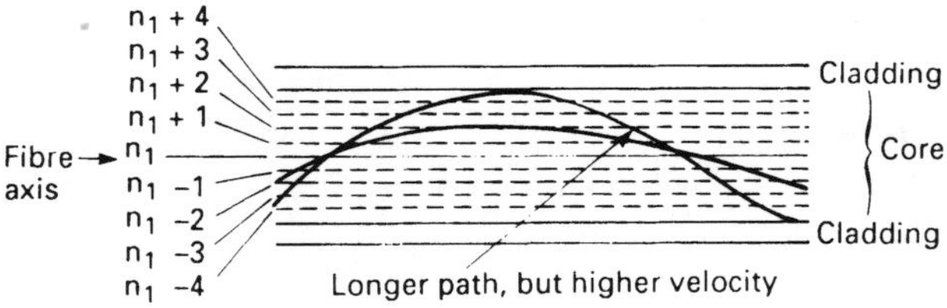

Table 3.1. *The three main optical fibre types compared*

Fibre type	*Bandwidth efficiency*	*Coupling diameter*	*Core*	*Typical cost* m^{-1} *(Jan 1989)*
Step-index	20 MHz km^{-1}	70	100–200	£1 ($1·7)
Graded-index	1 GHz km^{-1}	40	50	£1 ($1·7)
Single-mode	20 to 200 GHz km^{-1}	20	10	£5 ($8·5)

than axial rays. All light components therefore take almost the same amount of time to travel the length of the fibre, thus minimising dispersion losses.

Light is therefore transmitted along a fibre in a multitude of different paths, varying from those parallel to the fibre axis to those approaching the critical angle. Each path at a different angle is termed a **transmission mode**, and the numerical aperture of graded-index fibre is defined as the maximum value of acceptance angle at the fibre axis.

Typical attenuation coefficients of graded-index fibres at 850 nm are 2.5 to 3 dB km^{-1}, while at 1300 nm they are 1.0 to 1.5 dB km^{-1}, with a 3 dB bandwidth level of about 2000 MHz [13].

The main advantage of using graded-index fibre is the reduced refractive index at the centre of the core. This is particularly advantageous when trying to couple the tiny emitting area and sharply focussed beam of a small LED or laser diode to a fibre. Another advantage of using graded-index fibre is that it is comparatively cheap to produce, typically £1.00 ($1.7) per metre at current prices, and with better manufacturing processes and customer demand, even this is gradually coming down in price.

Table 3.1 compares the three main fibre types: monomode, step-index and graded-index.

Manufacturing processes and designs of optical waveguides

The advantages associated with optical fibres have been

known for a long time, but it wasn't until the early 1970s, that Corning Glass Works was able to produce a usable fibre, with an attenuation of about $16\,dB\,km^{-1}$, that optical fibres really became a viable possibility. Nowadays, of course, attenuations of less than $1\,dB\,km^{-1}$ and in exceptional cases as low as $0.1\,dB\,km^{-1}$ are to be found.

Fibres can also be manufactured from transparent plastic and these offer the advantages of a larger diameter (for example, 1 mm), increased flexiblity and, because the ends can be cut using a hot razor blade, ease of termination. Unfortunately, owing to the material's high intrinsic loss, the use of plastic fibres is normally restricted to only a few metres and to environments protected from temperature extremes.

How then is an optical fibre manufactured? Basically there are three separate steps, preparation of the glass preform, forming the glass rod and fibre drawing.

Preparation of the glass preform

Natural **silicon dioxide** (crystal quartz or silica sand) is manufactured into a solid, bubble free, glass preform of pure silicon dioxide by removing all metallic chlorides, e.g. iron, through **fractional distillation**.

Forming the glass rod

The preform is made into a glass tube of the desired core refractive index by doping the silicon dioxide with germanium and phosphorous to increase the refractive index, or with boron and/or fluoride to reduce the refractive index. There are three main methods of manufacturing the glass rod, internal vapour deposition, external vapour deposition and vapour axial deposition.

Internal vapour deposition

Figure 3.5 shows the preparation of the preform by internal vapour deposition. Silicon dioxide powder (doped to provide the required refractive index) is deposited in

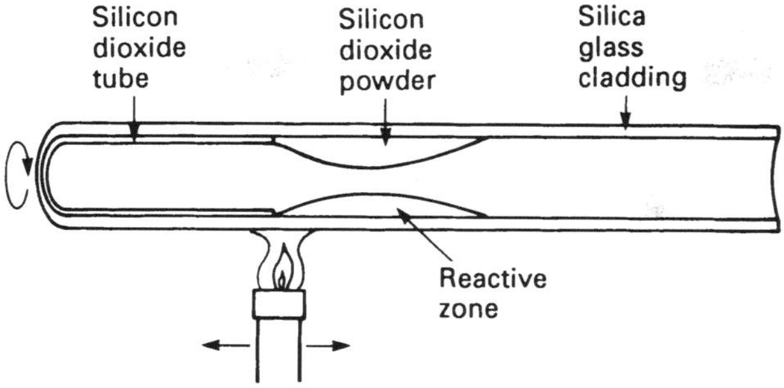

Figure 3.5 Glass preform preparation by internal vapour deposition

layers on a rotating tube. The tube is then heated to 1600°C to form core and cladding layers. By further heating, this hollow rod is collapsed to form a solid cylindrical rod with an internal core and a cladding of exactly the right refractive index. An outer cladding region (tube) of commercial silica glass then surrounds the rod. This process is referred to as the **deposition drying and sintering process**.

External vapour deposition

In the external vapour deposition process (Figure 3.6) layers of silicon dioxide are formed onto a ceramic **substrate rod** so as to make the required core and cladding properties. The ceramic rod is then removed and the hollow tube heated so that it collapses to form a solid rod.

Vapour axial deposition

A porous rod is heated, and variously doped silicon dioxide powders are deposited on it, *next to each other* (as

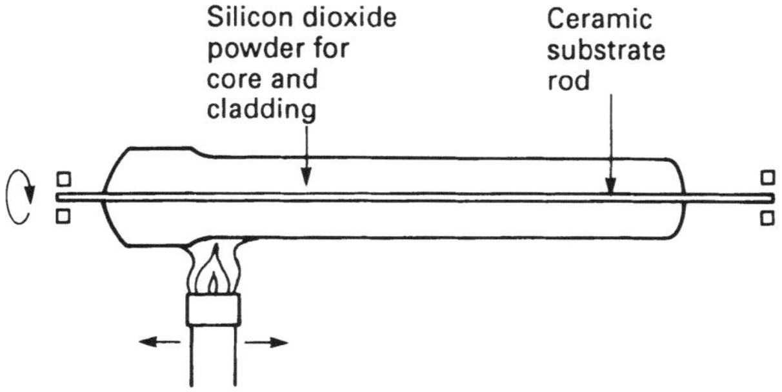

Figure 3.6 External vapour deposition, in which silicon dioxide layers are deposited onto a ceramic substrate rod

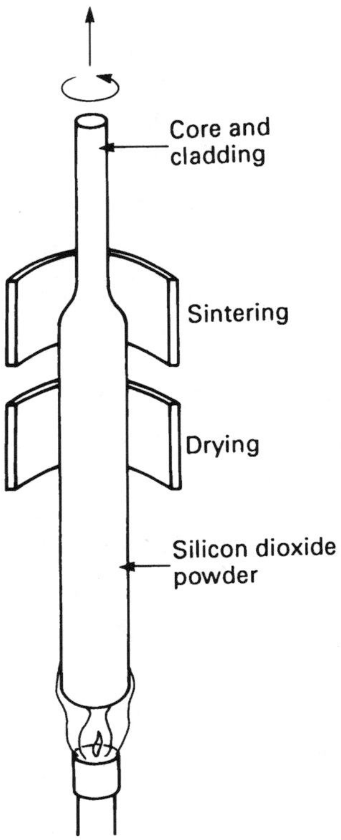

Figure 3.7
Vapour axial deposition of glass rod preform

opposed to after each other – as was the case in the other two methods). As it grows, the porous glass body is drawn off and heated to form a narrow solid rod. One advantage of this process is that the porous rod does not have to be removed. The area of the end face, however, limits the amount of silicon dioxide that can be deposited and therefore the growth rate. Figure 3.7 illustrates the process.

Fibre drawing

The glass rod obtained from one of the above methods is heated in a high temperature furnace to 2000°C and a thin fibre drawn off (Figure 3.8). It is important that this is done at a steady rate so as to minimise diameter variations in the fibre. Care must also be taken to ensure that the surface of the glass is not contaminated by foreign

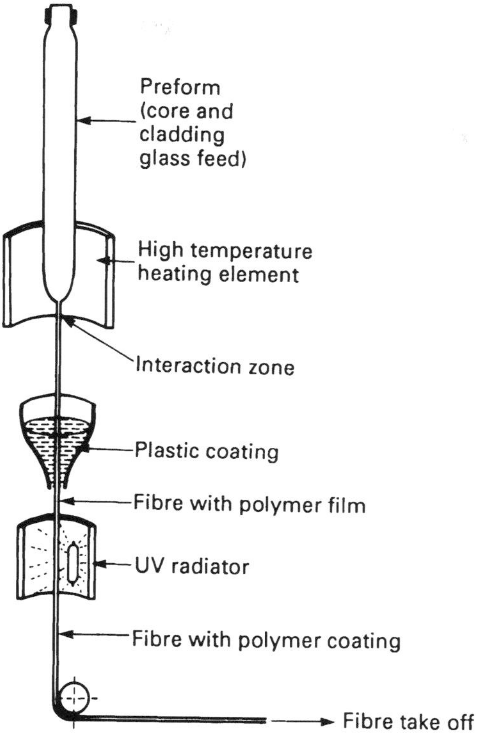

Figure 3.8
An illustration of the fibre drawing process

particles as this will cause **microcracks** in the fibre. Most manufactures nowadays look for a chemically **pure glass** with less than one foreign atom in 10^9 silicon atoms. A plastic cladding is now applied to the glass fibre. This makes fibres easier to handle, protects them from micro-bending losses and mechanical damage and is normally about 280 µm in external diameter. In special cases, for example, **light buffer cable** this outer coating could be as much as 500 µm in diameter.

So as to improve **fatigue resistance**, strength, resistance to **hydrogen-induced attenuation** increases and to protect against chemical corrosives such as hydrofluoric acid, a **hermetic coating** can be applied between the glass and the coating.

Other coatings, such as silicon and polytetrafluoro-ethylene (PTFE) (Teflon), can be applied so as to allow fibres to be used at high temperatures (up to 200°C), while metallic coatings can make fibre sensors respond more strongly to magnetic fields and pressures [13].

One of the problems of making a fibre in this manner is an undesirable dip in refractive index produced within the central core region. This is the direct result of the intense heat required for the collapsing and fibre drawing stages for heat causes the germanium dioxide core dopant to be evaporated away from the tube's inner surface. In multimode fibres, the refractive index dip can cause reduced coupling efficiency, reduced bandwidth and measurement errors. In single-fibre, the dip can result in increased **microcurvature sensitivity** and changes in the **cutoff wavelength** (ie, the shortest wavelength at which only the fundamental mode of an optical waveguide is capable of propagation – for single-mode fibres the cutoff wavelength must be smaller than the wavelength of the light to be transmitted).

In an attempt to overcome these problems AEG Telefunken has developed a far more efficient method of applying the dopant, called **full length outside deposition** (FLOD) [8] [18]. This process is shown in Figure 3.9.

In addition to maintaining a virtually constant refractive index throughout the fibre, the FLOD method has the advantage that it does not require such a high operating temperature. In the process, a tubular glass body with concentric regions containing a number of doped glass layers, each having a different refractive index is first produced. The glass body is then heated to its softening point and drawn into a glass fibre. During drawing, a partial vacuum is maintained in the hollow centre of the tubular body. The magnitude of this partial vacuum is sufficient to reduce the evaporation of the doping material from the tubular body interior, which effectively minimises the dip in the index of refraction in the centre of the optical fibre.

Cap sealing

To ease the problem of pulling an optical fibre through a

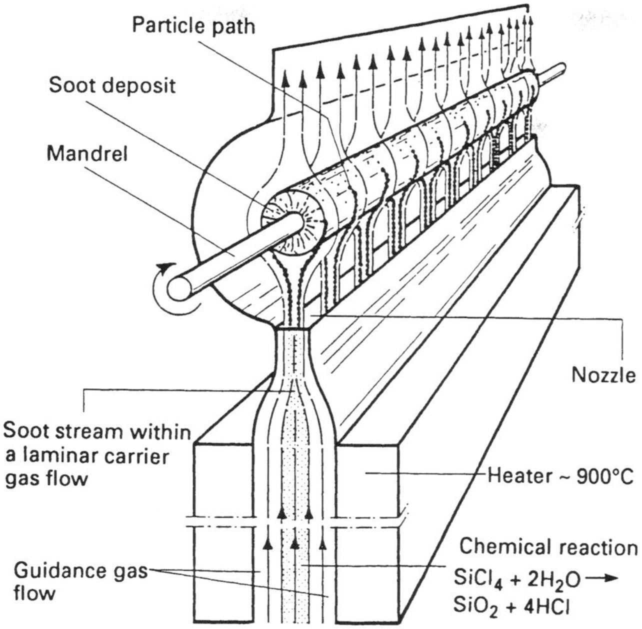

Figure 3.9
Full length outside deposition (FLOD) of glass rod preform

cable duct, fibres are normally provided with factory-fitted caps at both ends. Where this is not available, or short cuts of cable are being used, an 'eye' can easily be locally manufactured, as shown in Figure 3.10.

In small capacity optical fibre cables the factory-fitted cap is often transparent (shown in Figure 3.11) so that

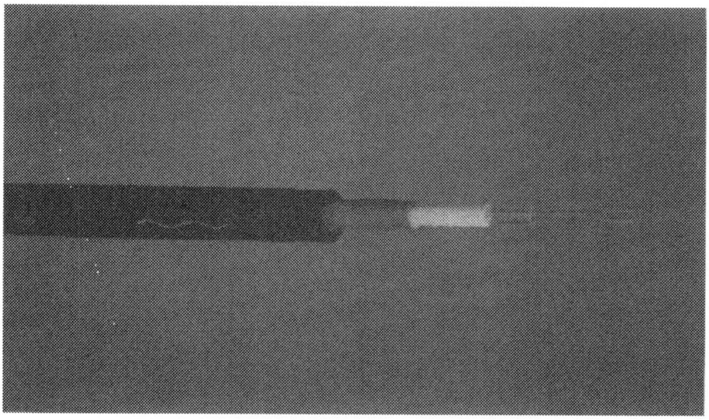

Plate 3.4
Ruggedised twin-fibre optical fibre cable (courtesy of Siemens)

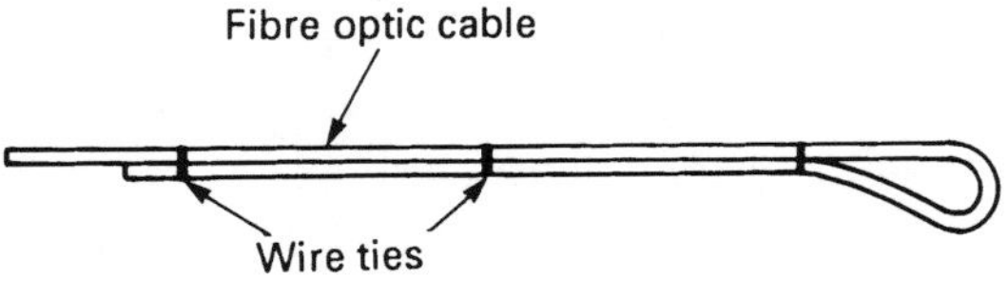

Figure 3.10 Optical fibre with hand-formed pulling 'eye'

the fibres inside can be illuminated to facilitate testing for damage after installation etc.

Cladding

There are two types of fibre designs currently available: one has a glass cladding; the other has a plastic cladding.

The advantages of **glass/glass fibre** is that a very clean **fracture surface** can be obtained which ensures that the fibre cladding inside the connector retains its optical characteristics right up to the end face of the fibre. With **glass/plastic fibre** on the other hand, some additional losses are introduced due to the fracture zone of the plastic which, even after grinding and polishing, will still have microscopic **end face absorption** areas.

British manufacturers

In addition to the well-known American (eg, Corning Glass Works) and European (eg, Siemens) manufacturers, British cable companies are also prominent in the industry, worldwide. For instance BICC, long renowed for its

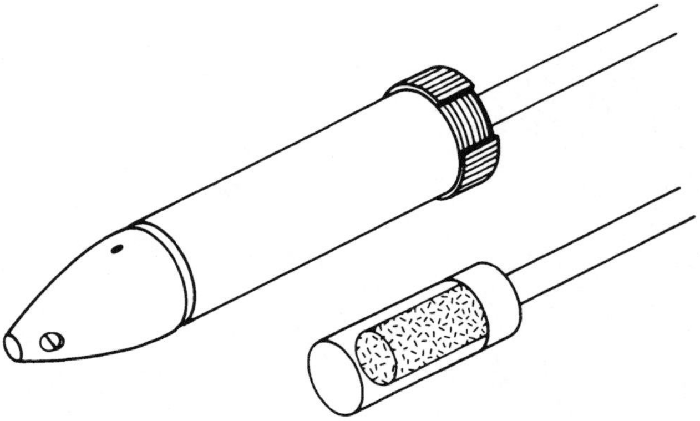

Figure 3.11 Transparent cap with pulling head

copper cable business, now possesses the largest optical fibre cable factory in Europe and STC can claim to be the largest British producer. STC in fact holds a market share of more than 60 per cent of the submarine cables that have been open to international competition.

Cable characteristics and specifications

Two of the most important characteristics associated with the installation of fibre optic cables are the bending and deflecting diameter. The **bending diameter** of an optical fibre cable is the minimum permissible diameter to which the cable can be bent without damaging the conductors. The **deflecting diameter** is the minimum permissible diameter over which the cable can be drawn or guided while being subjected to a high tensile force.

With an *all-dielectric* (ie, non-metallic) construction, cables can be bent to a comparatively small diameter without damaging the fibres. For example, cables with an outside diameter of up to 30 mm can be bent to 400 mm.

Cables with *metallic elements*, however, will have less favourable bending properties as they use **wire armouring** or have an **aluminum-laminated sheath** and copper conductors.

Tensile strength of the cable

Nowadays, most cables utilise high-tensile, low elongation elements of **aramid yarn**, encapsulated in an outer jacket with additional strength members. To meet the evergrowing requirement for heavy-duty cables, these **ruggedised cables** typically have a tensile strength in excess of 250 lbs.

Fibre diameter

The diameter of the fibre is an important consideration. Increasing the diameter of the fibre will reduce **tolerance problems** but the difficulty of producing a clean fracture surface is increased, **fibre flexibility** is reduced and costs are increased.

Optical fibre cable design

Optical fibre cables are naturally small in diameter and to enable them to withstand the bending twisting and pressure exerted on and to them during installation and whilst already laid, they must have some form of **protective shielding** around them. Various methods are available, including the following.

Slotted core cable

This is where a ruggedised cable is provided with **helical grooves** in which several fibres can be accommodated. The trouble with this type of cable is the difficulty in providing sufficient protection at junction points.

Buffered fibre

Possibly the most widely used method is buffered fibre. Here, the fibre is coated with a 60 μm thick protective sleeving such as polythene, ultra-violet cross-linked acrylate, or similar material and then placed either by itself (in which case it would look and act almost like a coaxial cable – Figure 3.12a) or together with a number of fibres (Figure 3.12b) inside a double-walled buffer tube.

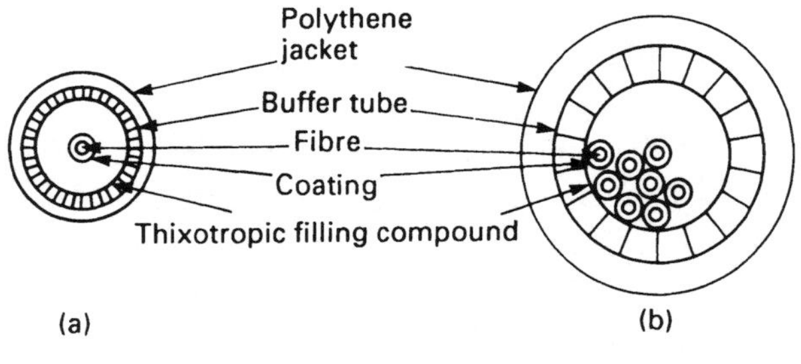

Figure 3.12 Construction of buffered optical fibre cable (a) single (b) multifibre

Normally, so as to enable easy identification, each fibre is colour coded with a 2 μm thin film of coating material. This coating has no effect on the optical properties of the fibre. Table 3.2 shows an example of the colour coding as used by Siemens.

The fibres are then encased in a polyisobutylene-based

Table 3.2 *Typical buffered fibre optical fibre colour coding, used by Siemens–*

Fibre number	*Colour*
1	blue
2	yellow
3	green
4	red
5	natural
6	blue (ring marked)
7	yellow (ring marked
8	green (ring marked)
9	red (ringed marked)
10	natural (ring marked)

filling compound to provide an additional buffer between the fibres and the buffer tube. These *thixotropic* filling compounds (gases, gels, pastes etc) cater for a wide range of environmental conditions, with temperature ranges of typically $-30°$ to $+70°C$. Because of this wide temperature range it is essential that the filling compound neither freezes, swells, nor runs.

The **flammability** of the buffer tube must also be insignificant and the tube must be capable of being easily cleaned. The tube and filling compound enable the core and fibres to adjust to a wide range of tensions and they are also impact resistant and laterally strengthened. Past experience has shown that the wall thickness of the buffer tube should be about 15 per cent of the overall buffer diameter and single-fibre buffers combined in one cable form should ideally have outside diameter of 1.4 mm while at ten-fibre buffer should have an outside diameter of approximately 3 mm.

The cable is often then encased in a rugged, flexible, age-resistant polythene (or similar) **jacket** which has a relatively low coefficient of expansion over a wide temperature range to allow the cable to be placed in ducts and pulled through in a similar manner to copper cables. A **high-tensile cap** (see Figure 3.11) can then also be connected to the support element.

Types of cable

Quite frequently a number of optical fibres are gathered

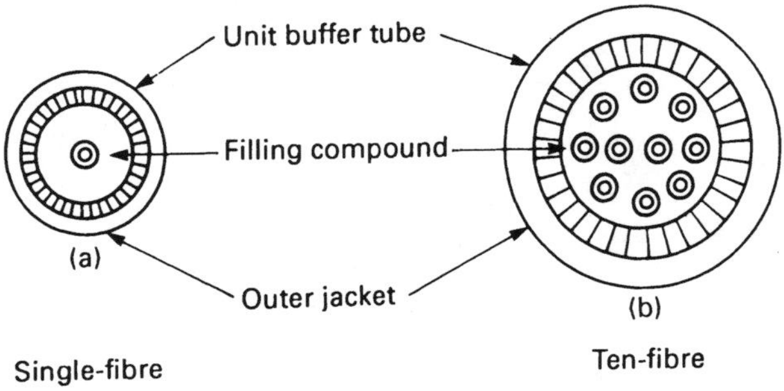

Figure 3.13 Construction of medium capacity optical fibre cables (a) single-fibre (b) ten-fibre

together in a random arrangement for use as a single transmission element, these are called **fibre bundles**. Fibres can of course be grouped together in a very similar manner to multipair copper cables and Figures 3.13 to 3.15 provide various examples of the construction and layout of some of the optical fibre cables currently available.

Medium capacity

Figure 3.13 illustrates medium capacity optical fibre cables. Figure 3.13a is a single-fibre cable, while Figure 3.13b is a ten-fibre cable.

Large capacity

A typical large capacity optical fibre cable is shown in Figure 3.14. In such **multiunit cable-forms**, fibres are

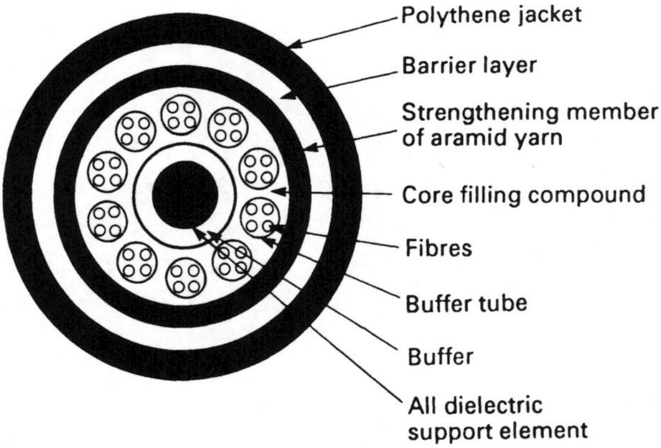

Figure 3.14 Construction of typical large capacity optical fibre cable

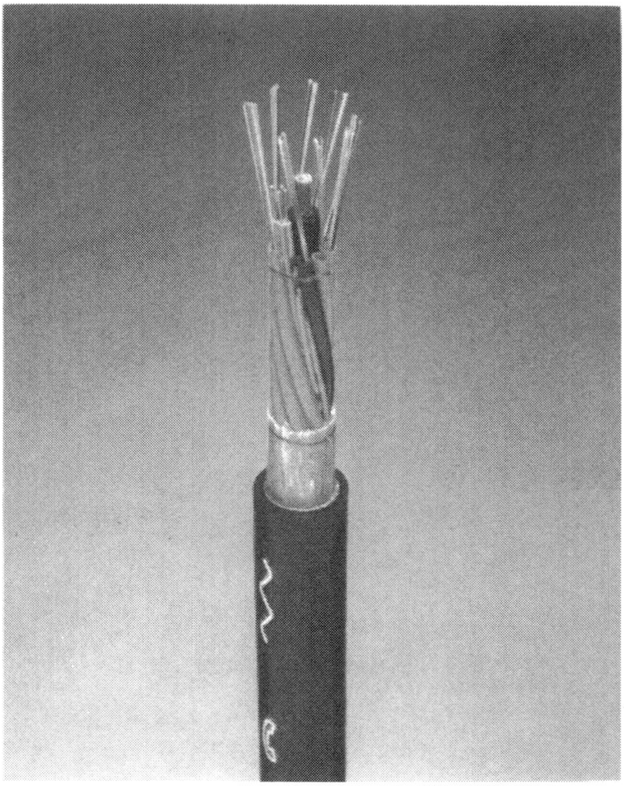

Plate 3.5
Ruggedised multi-fibre optical cable (courtesy of Siemens)

colour coded for identification purposes and sometimes more than one layer is used.

Ruggedisation of the cable is possible and this can be achieved by utilising some form of **strength member** (eg, **Kavlar**) which is either laid helically or braided around the fibre coating. This is then surrounded by a tough outer sheath to provide the required environmental and mechanical protection. Figures 3.15 and 3.16 show ruggedised optical fibre cables.

Installation of optical fibre cables

Training

The methods previously used in copper cables installation are identical to those used for optical fibres and the use of

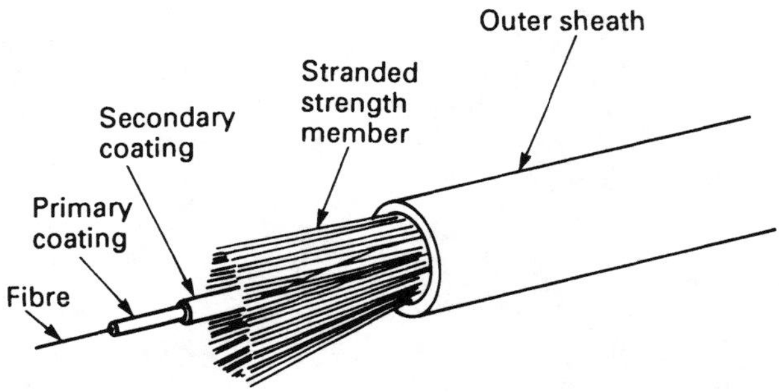

Figure 3.15 Construction of a single-fibre ruggedised optical fibre cables

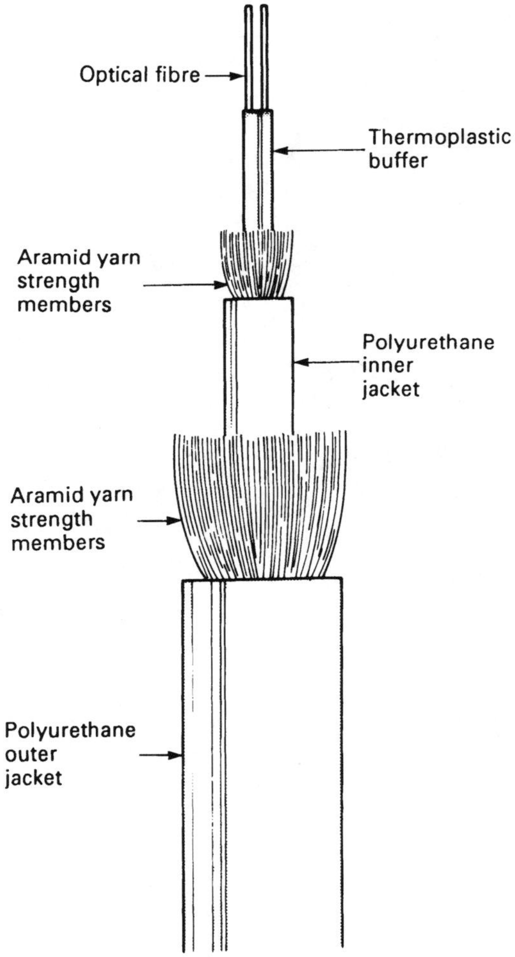

Figure 3.16 Construction of a multi-fibre ruggedised optical fibre cable, suitable for harsh, military environments

optical fibre cables requires very little additional or specific training except with respect to field maintenance (see Chapter 8).

Installation possibilities

Optical fibre cables, because of their mechanical strength and low weight, can be pulled into ducts, plowed in, cleated to walls, installed in vertical runs of over 1000 m in length, (this is particularly important in multi-storey office blocks and mines etc) placed on cable trays and planer shelves (or pvc conduit) installed aerially, attached to supporting wires, (eg, high voltage cables), or even immersed in water (oceanographic systems).

The technique used, of course, not only depends upon the environment but on the length of the cable to be installed. Today, one of the most practiced methods of installing short lengths into highly populated duct systems is to use compressed air to *inject* an **auxillary rope**. This will then be connected to a winch rope which in turn will be connected to the cable.

As the average pulling tension of one craftsman is only approximately 200N, often the preferred method of installing long lengths of cable (eg, 3 km) is to pull the cable in from the middle. Alternatively, **intermediary take-up points** are sometimes used.

Normally the first section is installed manually by two craftsmen. When the pulling tension approaches 200N (or at bends in the cable route) an intermediary take-up unit is installed in the last manhole passed by the cable. Another section is then pulled through until it is necessary to install another take-up unit. This method is repeated until the whole cable length has been installed.

A typical example of an intermediary take-up unit is the Siemens Unit [19], where the cable to be installed is fed over a wheel driven by an electrohydraulic motor. The pulling tension is applied in response to a small pilot tension caused through friction on the cable sheath. The speed of pulling in is controlled by the cable tip to

30 m min^{-1}. The system operates without oscillation or slippage and if the motor fails, the unit is allowed to free wheel. Corrugated sub-ducts are also often inserted into existing ducts and conduits, but because of their longitudinal elasticity, especially in heavily populated ducts, they tend to pucker in the direction of installation and thus increase the amount of friction. When this occurs, a non-corrugated plastic sub-duct should be used to provide a better installation performance. The utilisation of existing duct systems is of course possible and this can be achieved by using one of the previously mentioned methods, but to avoid jamming and wedging, the outside diameter of the cable should always be restricted to less than 5 mm.

Because optical fibre cables are very light they can be incorporated into overhead power line routes without any additional stress being exerted onto the the existing mast. In addition, as the optical fibre cable is completely non-metallic, any problems associated with inductive interference are eliminated.

The process is very similar to installing normal copper aerial cables and the optical fibre is merely lashed to the phase conductors or ground wires of the overhead power line (see Chapter 7).

4
Transmitters – light emitting diodes and lasers

Conversion of electrical energy into light waves

The conversion of electrical energy into optical energy is achieved by the use of an optoelectronic semiconductor device, such as a light emitting or laser diode.

Although the radiation from a light emitting diode (LED) is quite safe to the naked eye there is, of course, a potential hazard with the use of lasers and high-radiance LEDs because of the damage that can be caused through absorption of energy inside the retina. This is especially relevant as most optical fibre sources are, effectively, high intensity point sources to the eye. Because of this danger it is wise to supply operators, particularly those with constant access, with protective glasses and insist, as a precaution, that no energised optical source or illuminated fibre is viewed through a microscope.

The main requirements for lasers and LED devices is that they should be small, rugged, possess a long operating life, (at least comparable to other components within the system), have an emission only in the low attenuation spectral region of the optical waveguide, have a high coupling efficiency between semiconductor and optical waveguide, have a large radiating power (ie, high electrical/optical conversion efficiency), be capable of simple modulation by a transmission signal, have an economical construction and have low power consumption.

Light emitting diodes

With certain reservations silicon and germanium compound semiconductors meet these requirements, but as they both have a low breakdown voltage (caused by low electron mobility) and they are very brittle, more and more use is being made nowadays of gallium arsenide which, although more expensive to manufacture, has the advantage that it has a high breakdown voltage, high electron mobility, is a high speed device and so gives a relatively greater bandwidth.

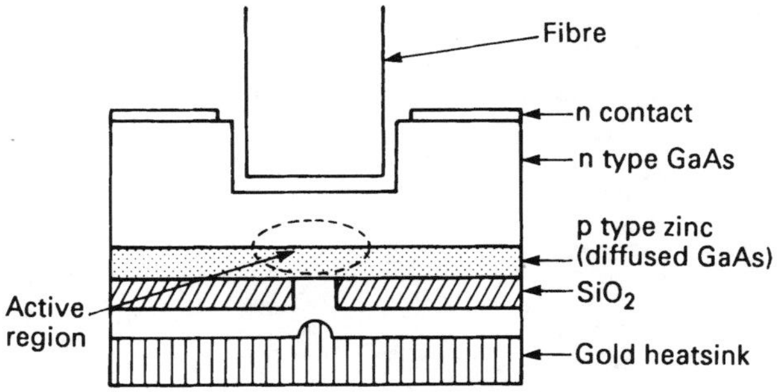

Figure 4.1
Small-area, high-radiance, gallium arsenide homojunction LED

The structure of a small-area, high-radiance gallium arsenide (GaAs) LED is shown in Figure 4.1. A light emitting diode is basically a semiconductor diode that is extremely efficient in converting electrical energy into light in a process known as **spontaneous emission.** It should be noted that as there is no optical feedback, an LED only produces **incoherent light.**

There are various disadvantages associated with using an LED as a transmitter. For instance, electrons can recombine without emitting light, they lose energy because they collide with other electrons (the **Auger Effect**) and the high refractive index of the semiconductor causes refraction and partial refraction of light at the semiconductor air interface.

As the average lifetime of a radiative recombination is only a few nanoseconds, the **modulation bandwidth** of an LED is limited to only a few hundred megahertz as opposed to laser diodes which can be modulated to many gigahertz.

Another disadvantage of LEDs is their **low coupling efficency** and the large amount of inherent chromatic dispersion. On the other hand, the main advantage of using LEDs in optical fibre systems is that their small core diameter permits an enhanced coupling efficency of the light into the optical waveguide. With a **spectral width** of about 40 nm, an LED has a bandwidth of at least 100 MHz and about 50 μW of **transmitted optical power** launched into the fibre.

Other advantages include simple design (and therefore ease of manufacture), simple system integration, low cost (between £10 to £100 ($18 to $180) at present) and high reliability. Additionally, as the light and current characteristics of LEDs are only slightly dependent upon aging and temperature, the design of an LED optical transmitter is comparatively simple and virtually only consists of a driver to modulate the LED.

LED preparation

For years, LEDs, like most transistors, were manufactured by the simple process of diffusing an element onto a crystal substrate. With the development of growth techniques this practice is now almost non-existent and the following are the preferred methods:

- **liquid phase epitaxi**: where extreme heat, (as much as 800°C in the case of gallium aluminum arsenide) is used to melt the source slice to provide a crystal deposit on the substrate
- **vapour phase epitaxi**: elements of the crystal are passed over a heated substrate and deposited epitaxially
- **molecular beam epitaxi**: the material is transferred directly as an ion beam from a heated source to a coated crystal substrate.

Types of LED

Edge emitting diodes

The manufacturing technique of **edge emitting diodes (ELEDs)** ensures that the inner p–n junction has a low

Figure 4.2
Edge emitting LED structure

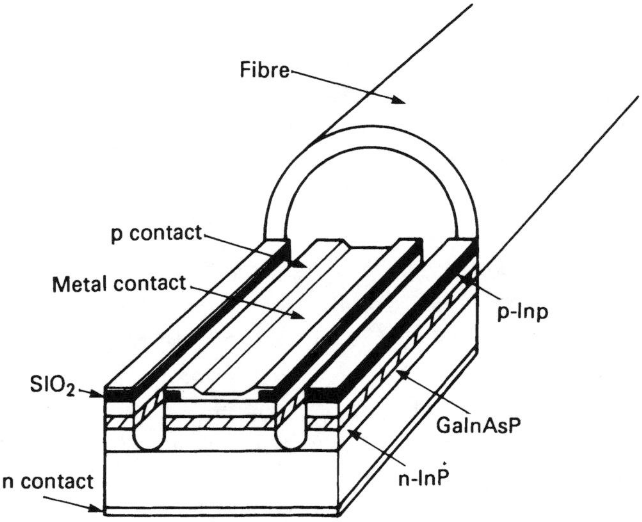

energy band gap whereas the outer regions have a relatively wide band gap. This **double heterostructure** make-up restricts the production of photons to a narrow concentrated beam because, although they might have sufficient energy to drift across the inner forward biased junction, they cannot overcome the higher barrier potential associated with the outer regions. The structure of an edge emitting LED is shown in Figure 4.2

Lateral spread is restricted by having a narrow width stripe electrode structure. In a similar manner to that of the fibre cladding, the outer layers also have a lower refractive index which helps to confine the emitted photons by internal reflection.

A new ELED (released in August 1987) is now capable of delivering 30 μW into a single-mode fibre. This far exceeds the 5 μW or so normally expected with most LEDs.

Edge emitting diodes have a comparatively thick active region which produces low self-absorption and an edge radiance far greater than the radiance viewed perpendicular to the junction.

By varying the mixture of the junction, different

energy gaps can be produced which provide different beam concentrations.

High-radiance light emitting diodes

Although LEDs operating in the visible red region (eg, 900 nm) are suitable for simple applications, systems operating in the infra red spectral region can provide transmission rates varying from a few Mbit s^{-1} to more than 100 Mbit s^{-1} over repeater distances of 15 to 20 km or more. High radiance LEDs, emitting at 850 and 1300 nm, are often used as they have an intrinsically high **quantum efficiency** due to the advantages gained by the double heterostructure of the diode. By designing the diode so that the size of the emitting surface layer is very limited, the **emission wavelength** can be set to low fibre attenuation and dispersion. The rise and fall are consequently very small and this method ensures that a low thermal resistance is maintained and a high radiance (therefore a high radiant power) is coupled to the fibre.

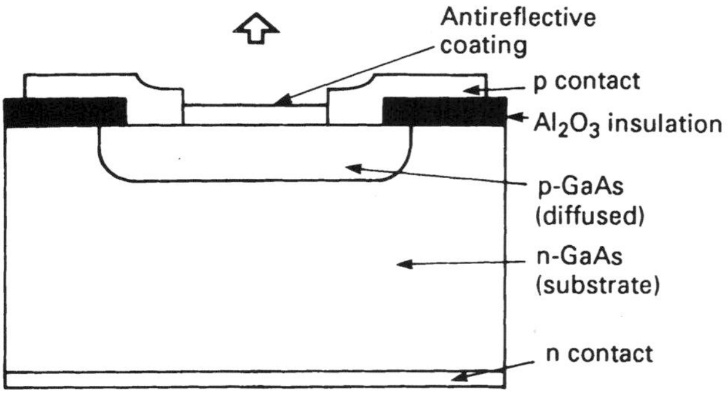

Figure 4.3
Structure of a diffused gallium arsenide LED

Diffused gallium arsenide LEDs emitting at 900 nm

This particular type of diode utilises planer technology and a typical example would be constructed as shown in Figure 4.3. The low **dislocation density** (less than 4000 cm^{-2}) of the n-doped substrate wafer minimises losses due to non-radiative recombinations at crystal defects.

The p-region acts as a **diffusion mask** and the p–n junction is graded to reduce **lattice strains** and additional losses due to, for example, crystal defects. To improve the radiation output of the diode, the final p-surface is coated with an anti-reflecting layer of silicon nitrate.

The emission wavelength of this particular type of diode would normally be about 900 nm, the spectral bandwidth approximately 40 nm, and the radiant power would be roughly proportional to the forward current (typically 120 μW at 100 mA).

The average lifetime (ie, time until the radiant power is reduced by 50 per cent) of one of these types of LED is approximately 10^5 h and diffused GaAs LEDs are suitable for transmission rates up to about 5 Mbit s^{-1}. Doping the p-region to a higher level, although facilitating large modulation bandwidths would reduce the amount of radiant power available.

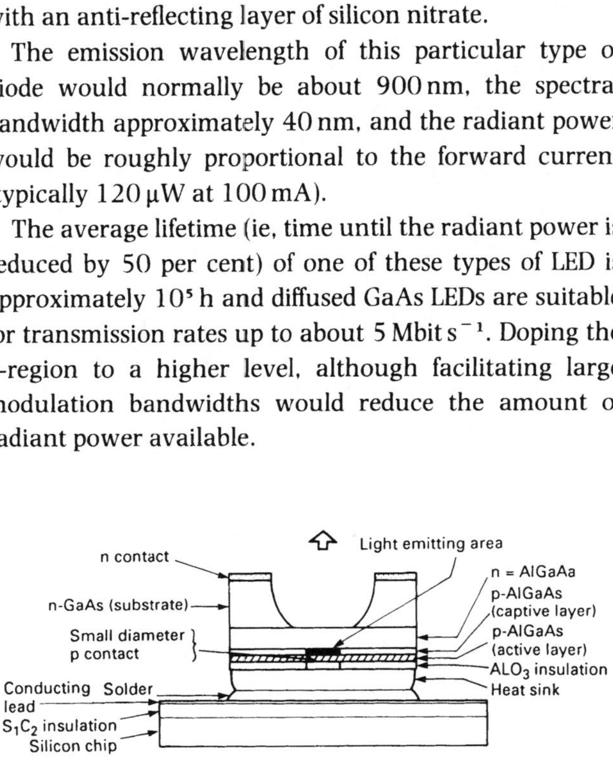

Figure 4.4
Structure of an AlGaAs/GaAs high-radiance LED

Aluminium gallium arsenide/gallium arsenide (AlGaAs/GaAs) High-radiance LEDs emitting at 830 nm

In this type of LED, shown structurally in Figure 4.4, three AlGaAs layers of varying thicknesses and doping levels (carefully selected to prevent any appreciable current spreading within the layer) are grown epitaxially on the GaAs substrate. To ensure that the AlGaAs layers have an extremely small transmitting area, the epitaxial layers point downwards. This is sometimes known as the **upside-down configuration**. The diode is mounted on a thin gold heatsink and soldered to a silicon chip containing a conductor track and insulation layer. Thermal resistance is minimised by locating the p-contact and the p–n junction near the heatsink. The high aluminium content of the two AlGaAs layers ensures that a potential barrier is formed to reduce the amount of electrons and holes that would otherwise be capable of entering the active region.

Quantum efficiency depends not only on the **crystalline quality** but also, to a significant degree, on the doping and composition of the active layer. Optimum values are achieved with compounds for wavelengths between 800 and 890 nm and doping levels around $2 \times 10^{18}/cm^3$. Such a diode would exhibit a rise time of approximately 15 ns [20]. By using AlGaAs layers the possible reduction of quantum efficiency due to non radiative recombination is minimised. The diode is then encapsulated within a very thin protective film and produces a spectral bandwidth of about 45 nm.

As the temperature rises the emission peak is displaced to longer wavelengths at the rate of 0·3 to 0·4 nm per degree Kelvin. Due to the small p-contact area, the diodes have a series resistance of 5 ohms. The voltage requirement for 100 mA is about 1.9 V and this produces a rise time of 15 ns. Shorter rise times (down to about 4 ns) are easily realised but only with a disproportionate decrease in optical power. It has been stated [21], that if the temperature rises the powers decrease by 0·3 to 0·4 per cent per degree Kelvin.

At 100 mA (dependent upon temperature) an optical power of approximately 4 mW is produced which would launch approximately 60 μW of optical power into a graded-index fibre and 700 μW into fibres with large core areas (eg, 100 μm). The life expectancy of a typical AlGaAs/GaAs high-radiance light emitting diode is in excess of 10^6 h [21].

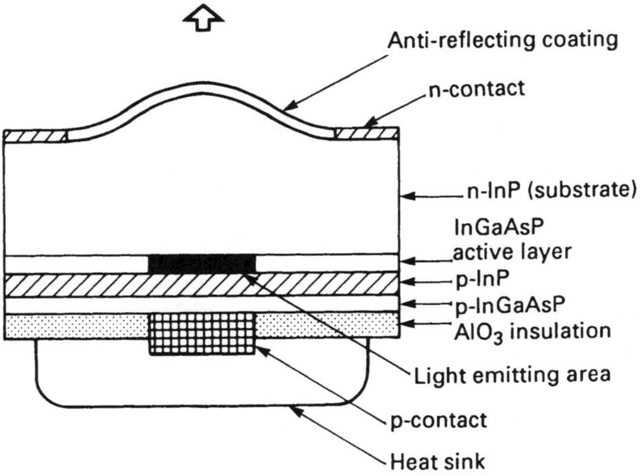

Figure 4.5 InGaAsp/InP high-radiance LED structure

InGaAsp/InP High-radiance LEDs emitting at 1300 nm

Figure 4.5 is a typical **quartenary high-radiance** LED and has basically the same construction as those previously described. A p-doped layer of InGaAsp is lattice-matched to a InP layer. This acts as a barrier and ensures a low resistant p-contact, but obviously even a slight lattice mismatch will reduce efficency.

To reduce reflection, the diode is given a thin coating of silicon nitrate. At 1300 nm the substrate is transparent and so hole etching is not required.

This type of LED would expect to have a rise time of approximately 8 ns and a fall time of 18 ns thus making it suitable for 34 Mbit s^{-1} transmission rates. Rise and fall times can be reduced to 3 and 6 ns (for 140 Mbit s^{-1} transmission rates) by increasing the **doping level**, but of course there would be a corresponding reduction in radiated power.

Although this particular type of LED reacts to temperature changes rather more than the others previously described (0.7 per cent per degree) they have a very low **ageing rate** (typically (10^6 h).

Laser diodes

After a photon has been generated (see Chapter 2) one of

two things can happen to it. It can either be radiated from the semiconductor or it can be absorbed in the creation of a hole-electron pair. If a photon should happen to be incident on a free electron in the conduction band it can stimulate the electron to prematurely recombine and in doing so emit a photon which is **coherent** (ie, in step, in phase and monochromatic) with the incident photon. The power of two coherent photons is twice that of two non-coherent photons so, if the diode current is above a critical level then there is a chance that the two photons will stimulate further emission (before they are either radiated or absorbed) which causes a rapid increase in radiant power. This action is called **lasing.**

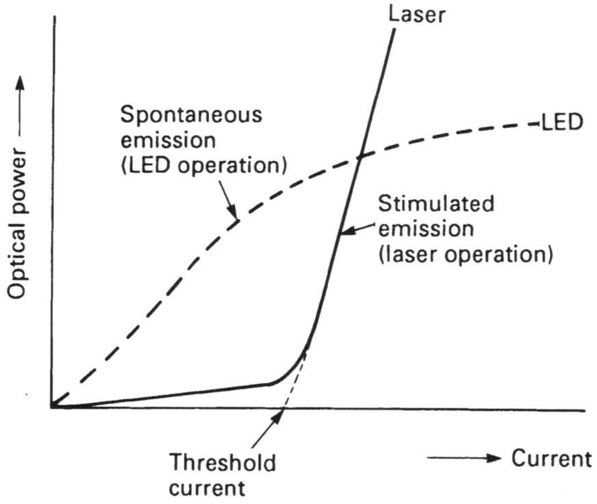

Figure 4.6 Optical characteristics of LED and laser diode compared

A laser diode is, therefore, essentially an LED with a wavelength selective element. Figure 4.6 shows a graph comparing optical powers of LED operation (due to spontaneous emission) and laser operation (due to stimulated emission), while Figure 4.7 shows spectral and spatial distribution differences between the two types of diode.

At low current the laser diode acts in a similar manner to a normal LED while above the threshold current **stimulated emission** (ie, the narrowing of the light ray to a few

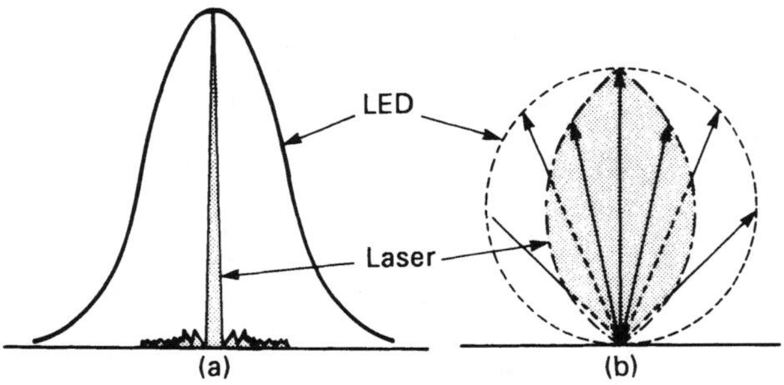

Figure 4.7 Comparing (a) spectral and (b) spatial distribution of laser diodes and LEDs

spectral lines instead of a broad spectral distribution) takes place. This not only enables the laser to be easily coupled to a single-mode fibre but also significantly reduces the amount of uncoupled light (ie, **spatial radiation distribution**).

Because the size of the threshold current is dependent upon temperature and ageing, the drive current must be regulated to provide a constant power output. To achieve this, light from the laser's back mirror enables the output of a photodiode to regulate the drive current.

Owing to their size, ruggedness and the need for only very low supply voltage to produce high speed modulation, the laser diode is a most cost-effective and efficient light source for long haul, high bit rate, communication systems. As a rule, at the lower end of the light spectrum (820 to 880 nm) gallium aluminum arsenide oxide stripe lasers are used, while laser diodes with their low fibre attenuation and a lower threshold current (eg, gallium indium arsenide potassium and indium potassium) are normally used at 1300 to 1600 nm.

If, in the double heterostructure ELED, the end faces of the emissive p–n region are cleaved and polished, sufficient photons would be reflected to maintain lasing action provided that the diode current is above the critical value. The region between two plane parallel, semitransparent crystal surfaces acts as a **Fabrey-Perot cavity** (ie, it performs as a variable bandpass filter or **spectrometer)** and is similar in action to a **waveguide resonant**

cavity in that it has very high Q, resulting in a narrowing of the spectral width of the laser emission.

As previously stated, the principle characteristic of a laser diode is that the light rays are coherent: they are in phase, they travel in the same direction, and they are virtually the same wavelength. This ensures that the laser light maintains a high energy density and that the ray does not diverge by any significant amount. Being coherent sources, laser diodes are virtually always used with monomode fibres.

Although gas solid state lasers (eg helium, neon, argon etc.) *have* been used in the past because of their size, cost and low efficiency, they are usually considered unsuitable for optical fibre applications and nowadays **semiconductor injection lasers** are used for most of the long haul and high bandwidth applications.

Advantages

Although the cost of a laser diode can be as much as ten times that of a simple LED – around £100 to £1000 ($170 to $1700) at current prices – the advantages of using lasers instead of LEDs include: simple economical design; high optical power; the ability to work at increased temperatures; an improved modulation capability (bandwidths well into the GHz range); high coupling efficiency; low spectral width (3·5 nm); the ability to transmit optical output powers between 5 and 10 mW; and the ability to maintain the intrinsic laser characteristics over long periods.

In addition, the **linearity** of an optoelectronic transmission system is mainly determined by the light current characteristics of the transmitter [15] as opposed to the **non-linear distortion** caused by the receiver and transmission medium, (which, unless very high transmission powers are used, are almost negligible) and although an LED exhibits signs of saturation at the higher power levels, the operational laser diode over a similar power

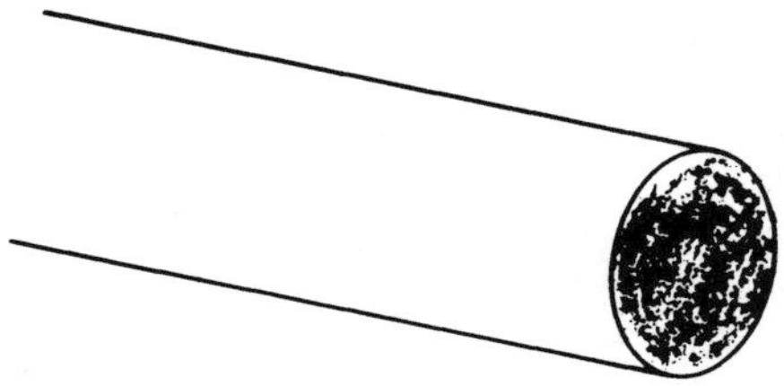

Figure 4.8
Two coherent light beams causing a speckle pattern at the end of an optical fibre

range, tends to improve as the power increases (see Figure 4.6).

Disadvantages

It should be noted however that, depending upon their relative phases, two coherent light beams can add or subtract their electric fields. This causes a **speckle pattern** at the end of the fibre as illustrated in Figure 4.8. Because of fewer transmission modes, graded-index fibre shows a more pronounced speckle pattern than step-index fibre. Although modulation rates are extended into the gigahertz range by the use of a DC bias (see later), one of the main disadvantages of using a laser diode is that it is extremely sensitive to overload currents and, at high transmission rates when the laser is required to operate continously, the use of a large drive current produces unfavourable thermal characteristics and necessitates the use of cooling and power stabilisation.

Because of the reduction in the number of electron carriers (caused by diffusion and equalisation of electrons and holes), the stimulated emission from a single p–n junction (as used in earlier laser diodes) only occurs at high threshold currents. In later models this has been overcome in a double heterostructure make-up, by sandwiching the active semiconductor layer between two semiconductor layers with a wider band gap. This structure prevents electrons injected from the n-side and holes injected from the p-side spreading outside the active region. Carrier confinement has thus been achieved. Various characteristics of a double heterostructure laser are illustrated in Figure 4.9.

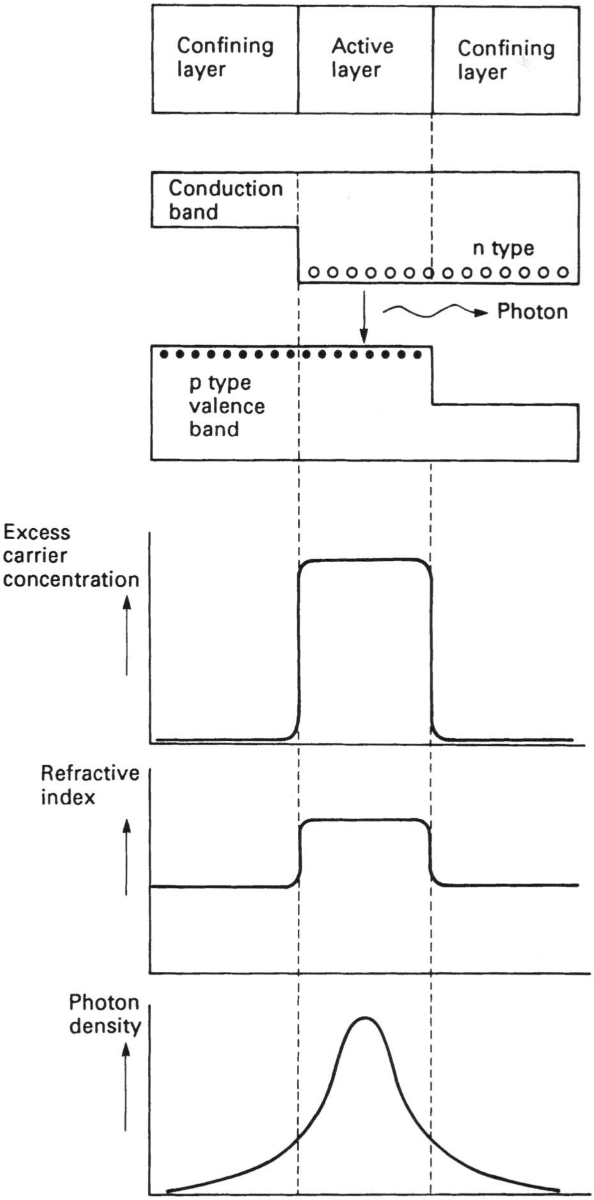

Figure 4.9 Advantages of a double heterostructure make-up

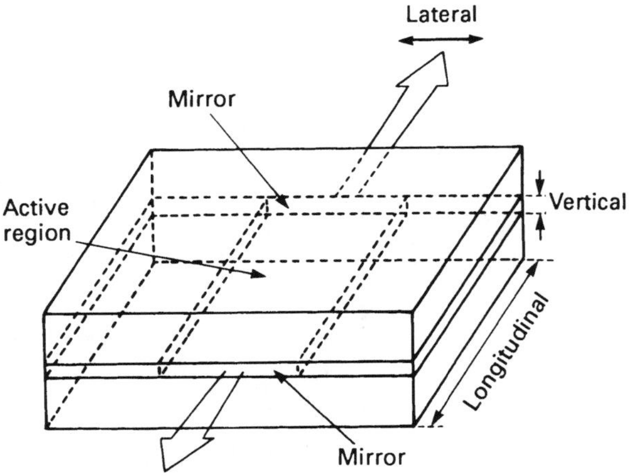

Figure 4.10 Spatial confinement of a laser diode's active region

As the wider band gap effectively has a smaller refractive index, sandwiching the active layer between two such layers produces a similar effect to that found in a glass fibre waveguide and the light wave is concentrated. The band gap (and hence the wavelength of the emitted light) can be varied by epitaxially depositing a new semiconductor on the substrate.

Design of a laser diode

Figure 4.10 shows the spatial confinement of a laser diode's active region. Longitudinal confinement of the electrical region around the p–n junction and the optical cavity is achieved by **cavity mirrors**; vertical confinement by the two mirror surfaces; and lateral confinement by additional internal structures.

Lateral confinement of the active region to a few microns is very important as this will restrict the lasers oscillation to the **fundamental transverse mode** and thus prevent high order modes being propagated. There are obviously quite a number of different types of lasers on the market today, but the most commonly used are those that depend upon the type of **lateral waveguide** that is used. There are two main types in this category: laser

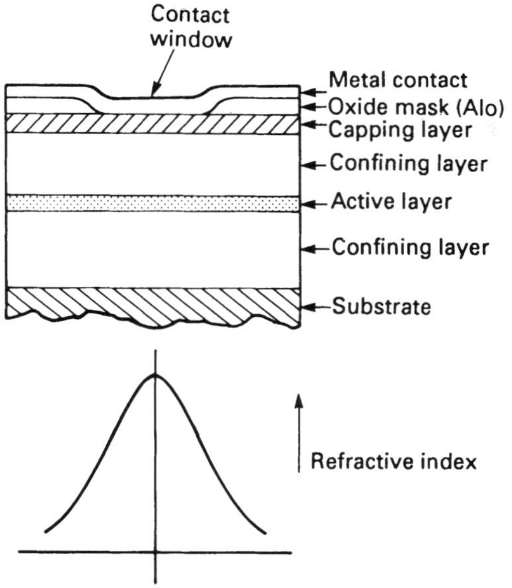

Figure 4.11
Oxide stripe laser diode structure

diodes with a current-induced waveguide; laser diodes with a built-in waveguide.

Laser diodes with a current-induced waveguide

Figure 4.11 shows a typical laser diode with a current-induced waveguide known as an **oxide stripe laser**. In a oxide stripe laser [22], the current carriers are concentrated onto an active stripe-shaped window which causes an **imaginary refractive index** to assume a lateral profile. As this lateral profile corresponds to an optical gain profile they are usually referred to as **gain-guided lasers diodes** (GLDs). Other types of GLDs are **proton-isolated lasers** and **v-groove lasers.**

Laser Diodes with a built-in waveguide

Examples in this particular group are **buried heterostructure** (BH) lasers, **channeled substrate planer** (CSP) lasers and **metal clad ridge waveguide** (MCRW) lasers. The MCRW laser structure is shown in Figure 4.12.

Lateral current confinement and waveguiding are achieved simultaneously, owing to the blocking effect of

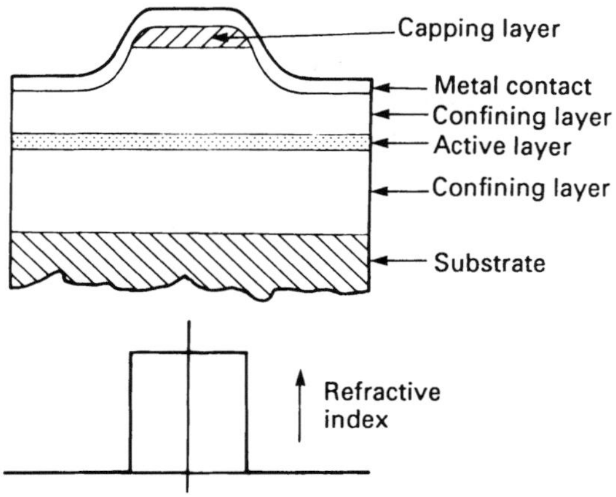

Figure 4.12 Metal clad ridge waveguide (MCRW) laser diode structure

the metal contact on the lightly doped capping layer outside the stripe. The increased reflection of the optical wave from the metallisation on both sides of the **mesa stripe** causes an effective refractive index and hence an optical waveguide. As all of this particular family have a built-in refractive index profile, they are also known as **index-guided laser diodes** (ILDs) [23].

Basic differences between gain-guided (current-induced) and index-guided (built-in waveguide) laser diodes

There are a number of basic differences between gain-guided and index-guided laser diodes. These include:

- **threshold current**: as the threshold current for an index-guided laser diode is approximately 20 per cent of that of a gain-guided diode, the amount of heat produced in an index-guided laser is much lower than a gain-guided laser
- **spectral width**: ILDs normally only emit one dominant spectral line at the peak of the gain curve. This characteristic is not maintained at high modulation rates, however, as the spectral position of maximum gain varies with modulation. The dynamic spectrum therefore broadens and the number of oscillating modes rises with transmission rate

- **temperature**: temperature affects laser diodes enormously. In 850 nm lasers the differential gain is lowered by typically 0.8 per cent per degree and the **threshold current** (ie, **lasing point**) rises by 1 per cent per degree. In 1300 nm lasers, threshold current rises 2 per cent per degree. Temperature dependence of the threshold current is expressed in terms of the **characteristic temperature.** Increases in temperature particularly affect ILD and GLD lasers, causing them to shift to a longer wavelength by as much as several tenths of a nanometer per degree. In GLDs this would normally only cause an **envelope shift** but in ILDs this could cause **mode jumping**
- **spatial radiation distribution:** the **far field** of a laser (ie, the spatial radiation distribution measured some distance away from the laser mirror) is affected by the size of the cavity (ie a **diffraction effect**) and is usually wider than conventional solid state lasers. Because of the differences in lateral and longitudinal dimensions the spatial radiation is normally asymmetrical
- **optical feedback sensitivity**: although maximum **launching efficiency** is desirable, the requirement to minimise the amount of reflected light fedback must also be considered. The **optical feedback sensitivity** varies widely between the different types of diode. ILDs, because of their high Q factor have a high sensitivity while a GLDs feedback is intrinsically lower.

Design characteristics

DC bias

Generally, a rapid applied current rise will only produce a stimulated emission after a delay – emission only occurs after the current has reached the threshold. This delay can be overcome by applying a DC bias current near that of the threshold.

Lattice matching

Reduced light output, ageing effects and indirect transi-

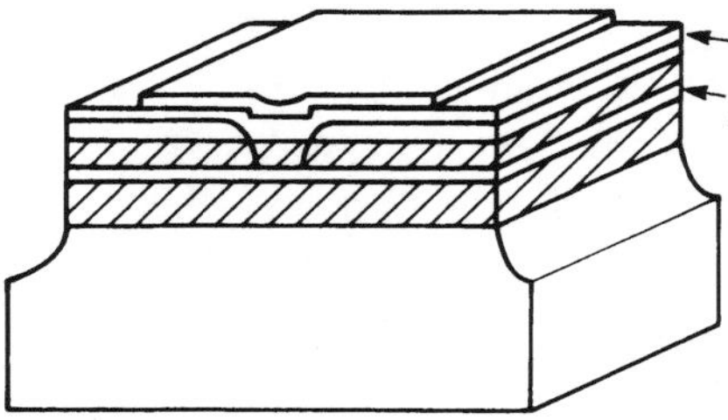

Figure 4.13 Structure of a diffused planer striped geometry laser diode

tions caused by lattice vibrations or crystal traps in the lattice [24] are important considerations. ILDs are generally more demanding.

Main types of laser diodes

Stripe geometry lasers

Stripe geometry lasers are gain-guided laser diodes. The principle behind a stripe geometry laser (Figure 4.13) is that the active width does not extend to the edge of the semiconductor chip and optical confinement is provided by the refractive index at the **heterojunction interfaces**.

Oxide stripe lasers

Oxide stripe lasers (of GaAlAs/GaAs construction) are used in optical multimode systems where the band gap of the GaAs determines the long wavelength limit and the Al content varies short wavelength limit.

The oxide stripe laser has a low sensitivity to temperature variations up to 100°C, so does not require cooling. In addition, oxide stripe lasers have an extremely stable laser emission even at high temperatures.

An essential part of an oxide stripe laser's highly stable format is its ability to handle high pulse powers and hence have a reduced sensitivity to spikes from electronic supply and regulating circuits.

Oxide stripe lasers are suitable for the transmission of large volumes of data over quite large distances.

Laser diodes in the 1300 to 1600 nm region

Because the fibre's attenuation is lower compared with

that at 850 nm, repeater spans are increased. Dispersion characterstics of the fibre are important and as a result, the previously mentioned dynamic spectrum broadening of lasers (due to the Fabrey-Perot cavity) becomes a limiting factor for transmission over long spans. However, a feature of the transmission band between 1300 and 1600 nm is that the overall dispersion of a single-mode fibre can be made to disappear almost completely at a particular wavelength thus negating the effect of **dynamic broadening** [25].

Although most laser diodes used for the long wavelength range are fabricated predominately from GaInAsP/InP, one of the problems associated with this type of diode is the increased amount of non-radiative recombination caused by increasing wavelength that limits the length of the wavelength.

As found in other communication methods, the phenomena of **phase interference** must be considered. Phase interference is caused when one light wave is superimposed on another at a different phase causing attenuation when antiphase and reinforcement when in phase.

In any particular fibre there will of course be hundreds of waveforms (called **eigenwaves** or fibre modes) that can be propagated without interfering with each other. The effect of delay differences between light components with different propagation paths is however the controlling factor with regard to the bandwidth of a fibre.

5
Receivers – photodiodes

Conversion of light waves into electrical energy

A **photodiode** is used to detect the optical light ray and convert the optical energy (**photons**) into electrical energy. Similar to LEDs and lasers, a photodiode's main requirement is that it should be small, rugged, possess a long operating life (at least comparable to other components within the system), be highly sensitive, possess a high response speed to the particular transmission rate (ie, be designed to detect the lowest possible signal powers at a given bit rate), have an economical construction and have a low power consumption. Similar to LED and laser transmitters, a semiconductor p–n junction can be used for optodetection. The principle of such a photodiode is illustrated in Figure 5.1.

In the depletion region the two charge carriers (holes and electrons) are separated by an electric field.

In the p–n diffusion regions the minority carriers dif-

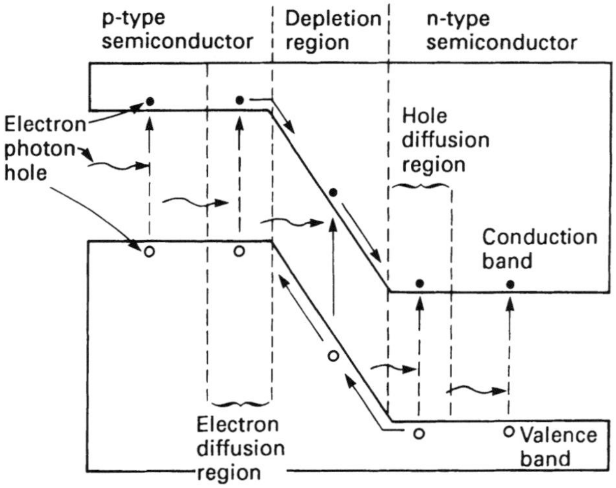

Figure 5.1 Basic photodiode operation

fuse to the p–n junction before they can reach the depletion region and be transported to the opposite neutral semiconductor region. This causes a **photocurrent** to flow in the external circuit.

Carriers generated outside the diffusion region are minimised because of recombinations and do not contribute significantly to the photocurrent. Because of this conversion efficiency is reduced.

The frequency response of a photodiode detector depends upon the **energy band gap** material from which the detector is manufactured and must be matched to the waveband of the light detected. The **quantum efficiency** of a photodiode is defined as the ratio of the number of generated electron-hole pairs to the number of incident electrons.

There are two main types of photodiode, the PIN and the avalanche photodiode.

Design characteristics of an optical receiver

Although the overall design of an optical receiver is fairly simple (consisting of only a photodiode to produce the current, and an amplifier), the design criteria of photodiodes is far more involved. This is because it must take into consideration a number of factors such as the noise from the signal current, noise from the photodiode itself (eg, avalanche diodes produce inherent noise dependent upon the multiplication factor), noise from the amplifier electronics, **receiver thermal noise**, **dark current noise**, as well as the signal-to-noise ratios required for the bit rate specified.

The PIN photodiode

The PIN photodiode is a reverse biased diode that has an **intrinsic** (ie, neutral) layer included between the n- and p-layers. The inclusion of the intrinsic layer decreases the junction capacity, increases the depletion region cross-sectional area, the maximum switching speed and the

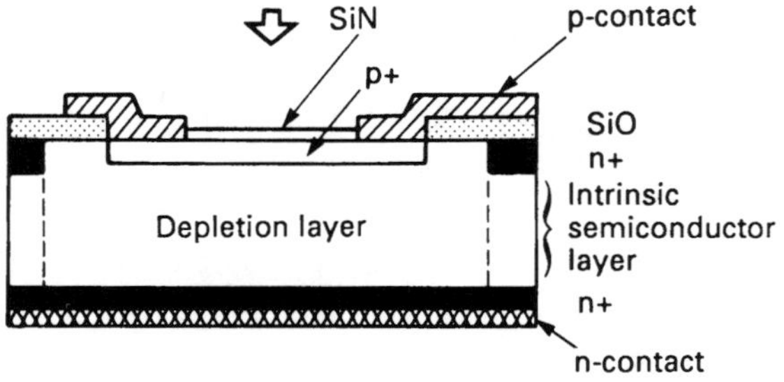

Figure 5.2 Structure of a silicon PIN photodiode

photon capture area. A PIN photodiode's structure is shown in Figure 5.2

In the absence of incident light, any **reverse current** is due to thermally generated electron-hole pairs and is called **dark current.** If photons are incident on the depletion layer, they will also create electron hole pairs and the total reverse current will increase in proportion to the incident light intensity.

The main advantages of using PIN photodiodes is that they have good linearity, high bandwidth and good temperature stability.

Silicon is normally used when operating in the near infra-red region, InGaAsp in the long wavelength and germanium in both the infra-red and the long wavelength regions.

Avalanche photodiodes

By operating the photodiode in the avalanche photodiode (APD) or **breakdown mode**, the acceleration of carriers in the electric field reach such a high velocity that **impact ionisation** occurs and further electron-hole pairs are generated which internally amplify the photocurrent produced.

As the carrier transmit time increases, the gain-bandwidth product remains almost constant but large enough to convert optical powers in the nanowatt range at frequencies over a gigahertz.

For higher frequencies (and high input sensitivity), on

the other hand, the PIN diode would usually be followed by an electrical amplifier.

Avalanche photodiodes, although non-linear and also fairly unstable, are very similar to normal silicon photodiodes except that they require a slightly lower operating voltage to achieve good multiplication. A basic silicon photodiode would, for example, require several hundred volts to achieve the high field strength gained from avalanche multiplication.

A **guard ring** is normally used to prevent edge breakdown and to limit possible inversion layers at the surface. Silicon is mainly used for the near infra-red region while germanium is normally used in the long wavelength region.

Silicon photodiodes

Basic **silicon photodiodes** (because of their small capacitance and low dark current noise) are very suitable as optical receivers. However, due to their absorption rate being limited by the silicon band gap, silicon (and for that matter, silicon avalanche) photodiodes are normally restricted to use in the 800 to 900 nm range.

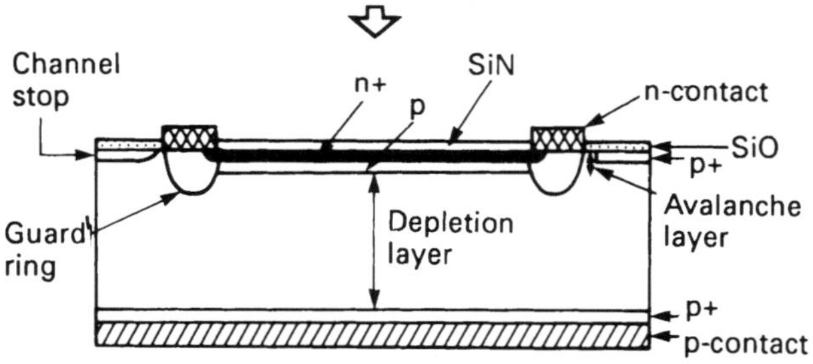

Figure 5.3 Silicon avalanche photodiode structure

Figure 5.3 shows the structure of a silicon avalanche photodiode. Light passes through the p-layer (which is protected by a thin coating of silicon nitrate to prevent reflection losses) and separation of carrier pairs is achieved in the depletion layer (electrons to the n-side and holes to the p-side).

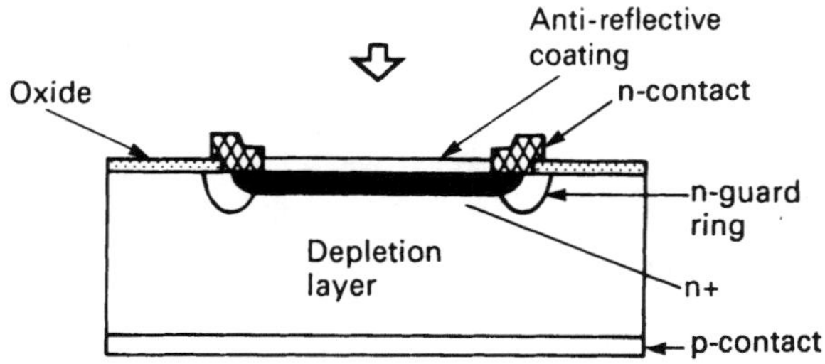

Figure 5.4 Germanium avalanche photodiode structure

Germanium photodiodes

Because of the silicon photodiode's low absorption rate above 1100 nm (caused by the silicon band gap) germanium and alloys of various similar compound semiconductors are used in the 1300 to 1500 nm range.

In the structure of a germanium avalanche photodiode shown in Figure 5.4, the diode consists of a lightly doped p-type germanium substrate surrounded by an n-doped guard ring with p-n junction produced on the surface by diffusion or ion implantation. An anti-reflective coating of, for example, silicon oxide increases the quantum efficiency of the diode.

Indium gallium arsenide photodiodes

Unfortunately, germanium photodiodes also have limitations in this spectral region such as high noise, temperature sensitivity, low responsiveness and dynamic impedance. Because of these problems indium gallium arsenide photodiodes are quite frequently used.

The main advantage of InGaAs photodiodes is that noise equivalent power is less than half that of the same size germanium diode. This enables the photodiode to detect at greater distances and with increased accuracy.

6

Waveguides, couplers, connectors and repeaters

Coupling optical fibre components and systems

When an optical fibre cable is damaged it can be repaired either by splicing a new piece of cable into the existing line or, if sufficient cable is available, by cutting out the damaged portion and splicing the two ends together. This process is referred to as **concatenation**.

Quite frequently, especially when it is necessary to attach filters, optical couplers or transmit/receive opto-electronic transducers etc to the fibre, a low-loss, easily demountable connector is used. There are two main types of **optoelectronic demountable couplers**: end fire couplers and lens couplers.

End fire coupling

End fire coupling is where both optical faces (whether they are fibre-to-fibre, fibre-to-photodiode, or LED/laser-to-fibre) are parallel and close to each other. Various coupling losses have to be considered, the most important being **insertion loss**.

The three main situations encountered in end fire coupling are shown in Figure 6.1. If the core diameter of the entrance and exit faces of the fibres are similar in size (Figure 6.1a) and they also have identical numerical apertures, this is called **symmetrical end fire coupling** and losses are minimised.

If the exit face is larger (Figure 6.1b) than the entrance face (eg, LED-to-fibre) high losses will occur.

If a small exit face (Figure 6.1c) is coupled to a large entrance face (eg, fibre-to-photodiode) the insertion loss is low.

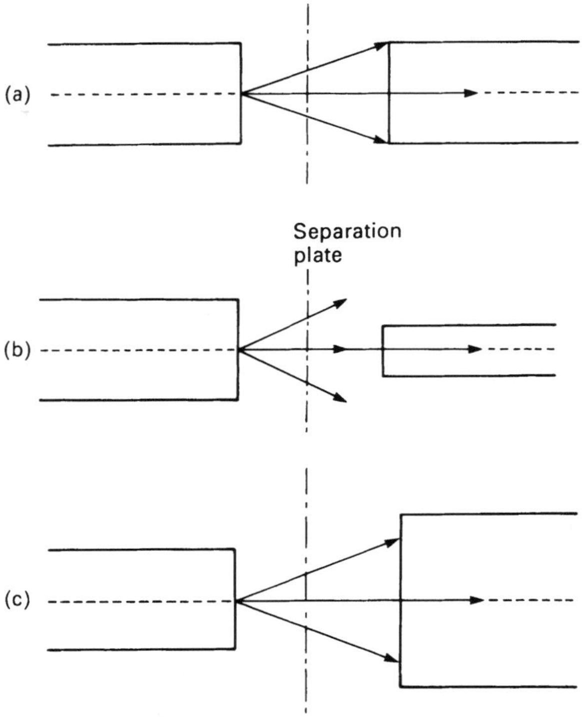

Figure 6.1
Three main situations encountered when end fire coupling optical components
(a) symmetrical end fire coupling; when entrance and exit faces of the components are similar in size
(b) when exit face is larger than entrance face
(c) when exit face is smaller than entrance face

Lens coupling

This particular method uses lenses to couple the exit face to the entrance face. Of the many examples, **symmetrical lens coupling** (Figure 6.2a – a fibre-to-fibre coupling) and **asymmetrical lens coupling** (Figure 6.2b) where the exit face is smaller than the entrance face (a fibre-to-photodiode coupling) are the most widely used.

Although lens coupling permits greater mechanical tolerances at the junctions, the losses of a lens coupler are usually much higher than that of an end fire coupler, because of the additional losses caused by the imaging system and the different refractive indexes of the bonding layers. For this reason, end fire coupling is normally the preferred method.

Coupling losses

Connector insertion loss is not just a function of the

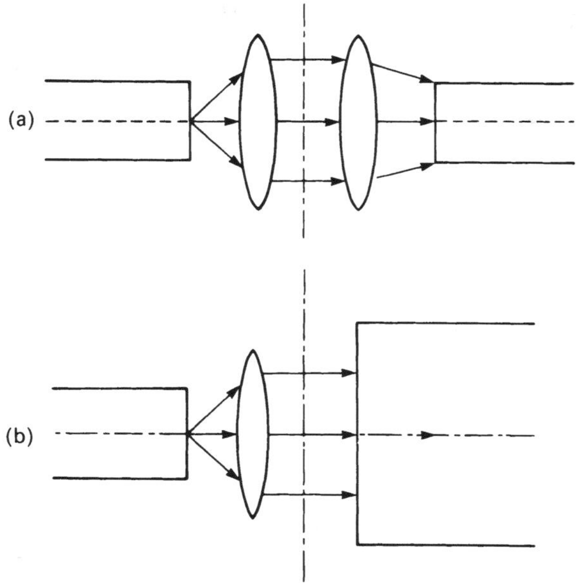

Figure 6.2
The two most widely used forms of lens coupling optical components
(a) symmetrical lens coupling: from fibre-to-fibre
(b) asymmetrical lens coupling: form fibre-to-photodiode

tolerance of the connector but also the tolerance of the fibre itself and there are a number of factors that must be taken into consideration when designing optical fibre systems:

- **axial fibre misalignment**: probably the most critical factor as it can introduce extremely high losses (Figure 6.3a). For example, an axial (transverse) offset of around 30 per cent (equivalent to 30 μm for a fibre with a core diameter of 100 μm) is sufficient in itself to introduce a loss of approximately 2 dB [26].
- **angular misalignment** (Figure 6.3b): not so important in large fibres but the smaller the numerical aperture, the more significant the problem
- **Fresnel losses**: (ie, those due to the reflection of the glass-air interface) are of particular importance and can be as much as 0.35 dB at fibre-air-fibre combinations and the separation between fibres (Figure 6.3c) is, therefore, another important consideration. Fresnel losses can, however, be reduced by using a dielectric coating and/or an index matching liquid.

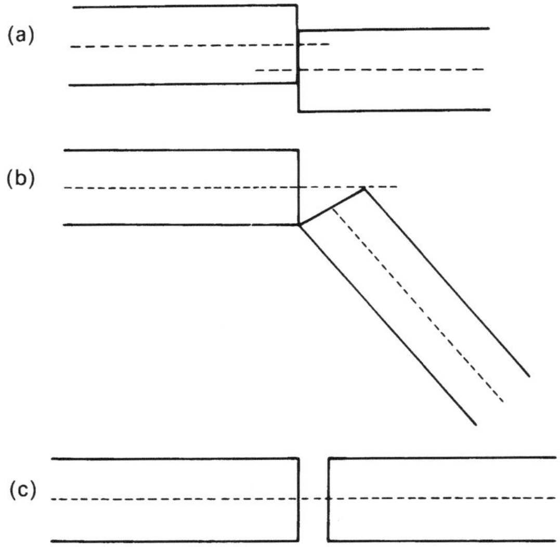

Figure 6.3 Connecting problems which can cause severe coupling losses (a) axial fibre misalignment (b) angular misalignment (c) separation between fibres

The actual surface finish, or **endface roughness**, of an exit or entrance face also affects the amount of insertion loss that a junction exhibits as it can cause **light scattering** and **absorption centres**. For example, a fibre endface with a peak valley height of 10 μm introduces a loss of around 0.5 dB [26].

Core diameter variations between the exit and entrance fibre core diameters also introduce insertion losses. For example, a core diameter variation (involving a transition to a smaller diameter fibre) of 5 per cent, will introduce a loss of around 0.5 [23].

Connectors

Single fibre connectors

Figure 6.4 shows an example of a graded-index fibre connector where the cable is secured to the plug by crimping. The cable can then be connected, or screwed, to an optoelectronic transducer using the coupling nut located on the plug.

Many companies support their fibre ranges with such connectors. For example, STC has recently produced a

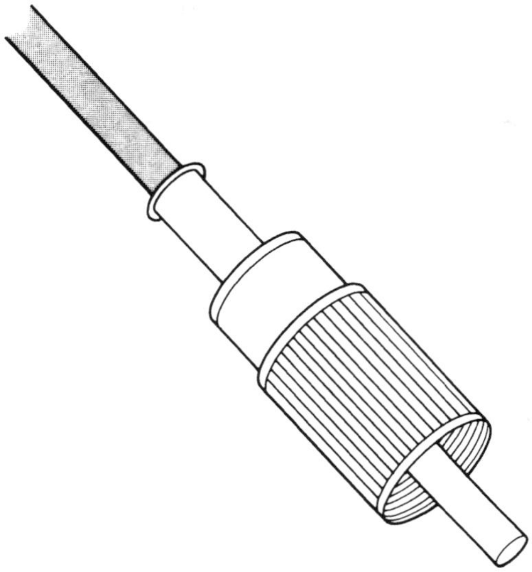

Figure 6.4
Single fibre connector

range of optical fibre cable assemblies designed specifically for **harsh environments**. These are composed of twin fibres, sheathed with high strength, low flammability, polyester, that have a breaking strain of 2000 N and are terminated with **hermaphroditic expanded beam connectors**. The cable and connectors have an overall loss per kilometre of less than 4 dB at 1300 nm. The weight of one kilometre of the cable is 31 kg so this particular cable is about a quarter the weight and half the size of the military coaxial cable it replaces. The connectors are shown in Figure 6.5.

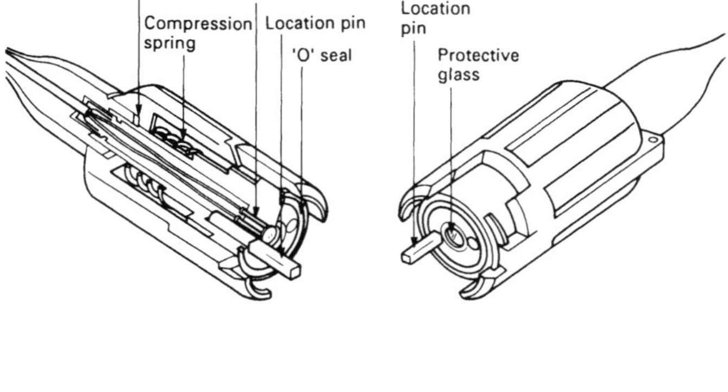

Figure 6.5
Optical fibre cable assembly manufactured by STC

Each connector is provided with a sealed bayonet locking cap attached by a stainless steel lanyard and each connector has a window to protect the lens and to facilitate easier cleaning. The windows and lenses are coated with an anti reflective film to reduce losses and crosstalk. In the event of a window becoming permanently marked it can be replaced in seconds using a simple hand tool. Additionally, the connectors and cable materials are all proofed against petrol and diesel fuels, lubricants, hydraulic oils and alkaline wash solutions.

The **bulkhead connector** is basically similar in construction to the free connector and the body can be moulded in carbon fibre loaded polycarbonate to maintain electromagnetic shielding across the panel cut out.

Connectors for fibre bundles

Connectors for fibre bundles (which can literally contain hundreds of single fibres) are readily available nowadays and although insertion losses can be as high as 3.5 dB, because of the high attenuation of fibre bundles (50 to 700 dB km^{-1}) they are relatively insignificant [26].

Attaching the fibre to the connector

This can be achieved by cleaving and either cleating the fibre to the ferrule so that the cleaved surface forms the fibre exit face, or using a low attenuation adhesive to secure the fibre to the ferrule.

The ability to consistently cleave to an accuracy of better than one degree every time is currently required for today's splicing and measurement applications. Fortunately there are a number of cleavers and splicers on the market at the moment that are capable of doing this.

Connecting an optical fibre cable to an LED

Although fibre **butt coupling** (Figure 6.6) is possible it is very inefficient owing to the relatively narrow acceptance angle of the fibre and so couplers using lenses to concentrate the beam (Figure 6.7) are more widely used.

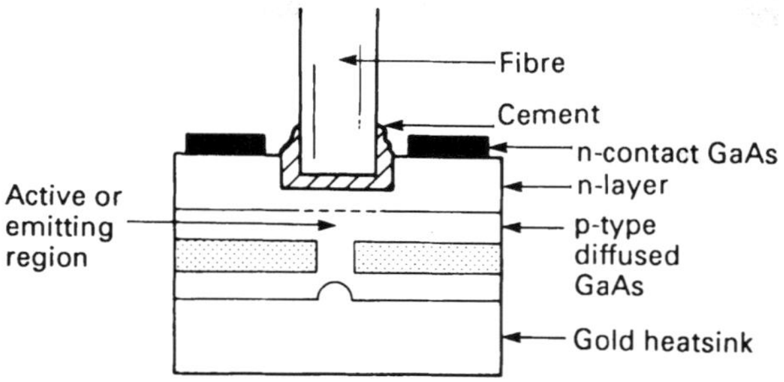

Figure 6.6
Butt coupling an optical fibre to an LED

Although, from a thermodynamic point of view, it is impossible to increase the radiance of an LED by using a lens coupler, lens coupling does, in fact, lightly magnify the active area and obviously increases coupling efficiency.

Connecting a fibre cable to an integrated circuit

There are two methods currently used to connect an integrated circuit to a fibre optic cable. They are end fire coupling and evanescent wave coupling.

End fire coupling

This is effectively a butt coupling of the **integrated optical waveguide** to the optical fibre which is then held together by an adhesive.

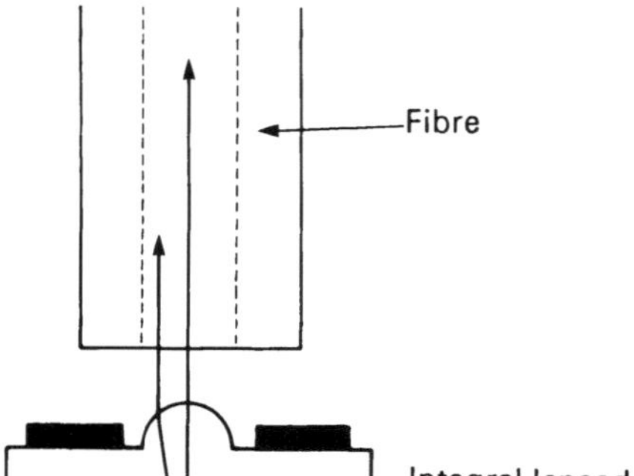

Figure 6.7
Lens coupling an optical fibre to an LED

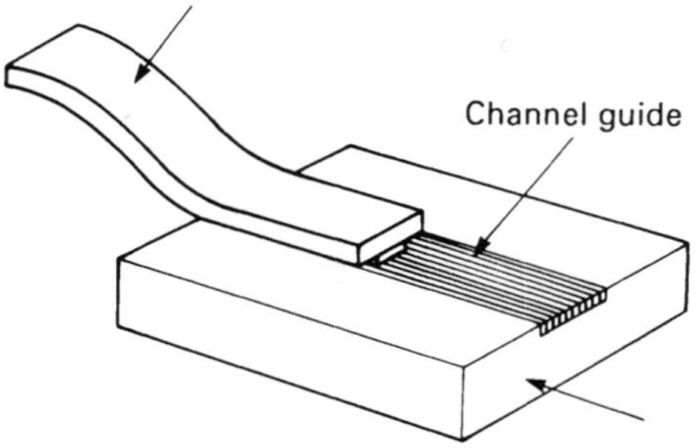

Figure 6.8 Coupling configuration for a sandwich ribbon fibre with an integrated waveguide

Evanescent wave coupling

In this case (Figure 6.8) a **sandwich ribbon fibre**, having its core on the fibre surface, is pressed against the integrated optical waveguide. Coupling efficiencies of over 90 per cent have been achieved using this method.

Quality of the connector

The quality of the connector depends upon a number of things such as the type of fibre diameter, fibre diameter tolerances, fibre asymmetry, concentricity of the fibre core, numerical aperture variations and the type of cladding used.

Requirements for fibre optic connectors

The basic requirement of a fibre optic connector is that it should be easy to assemble, have a rugged design, be low loss and have reproducible loss values after repeated matings. It is also very important that fibre optic connectors are so designed that they do not damage the actual fibre end and in so doing introduce more losses.

Fibre jointing techniques

Joints, or **splices**, unlike connectors, are permanent connections between fibres. As a rule, splices are normally used in external, long distance, high capacity optical fibre systems whilst connectors are used in internal short distance applications. This is particularly significant as the loss of a connector is typically 0.7 dB whilst that of a splice is usually 0.1 to 0.5 dB depending upon the type of splice and fibre used.

In a splice, the two fibre ends are bonded together either by melting (known as **fusing**) then gluing or, mechanically holding them together in a tightly confined structure. This process normally takes between 5 and 10 minutes to complete [22]. In general, mechanical and epoxy splices lack the strength of fusion splices.

Bonding the fibre ends together with no intervening air space reduces the Fresnel and transmission losses normally found at fibre-air interfaces. Some interface and spreading losses can occur if the fibre ends are separated by a bonding agent such as an **epoxy resin**, but these are minimal if the refractive index of the material is close to that of glass and the layer is not too thick.

It should be noted, however, that improper splicing can allow dirt into the junction and this will form a **lossy junction**. For example, a solitary 10 μm dust particle would almost totally block light transmission through a single-mode fibre.

Splice losses

Splice losses can be due to variations in the outer diameter of the fibre core, differences in index profile, differences in the ellipticity of the core, misalignment of the fibre ends, poor quality of the refractive index match at fibre ends, waveguide imperfections etc.

In practice, splice losses of about 0.5 dB are typical for fusion-spliced multimode fibres while single-mode fibres (because of the narrowness of the core diameter) are not as much (for example 0.1 to 0.2 dB). Mechanical splicing, while still being more lossy than fusion-splicing, is nevertheless appreciably less than that of a connector.

Although it is physically possible to splice non-similar fibres (eg, single-mode to multimode) sizeable losses can be expected if the light emerging from the output fibre is not completely collected within the core of the input fibre. For example, connecting a 10 µm core single-mode fibre to a 62.5/125 µm multimode fibre would be minimal, for a light beam in the single-mode to multimode direction, but in the other direction it could be as much as 20 dB.

As the loss due to a splice can sometimes be as much as the loss due to a kilometre of single-mode fibre it is important that the splicing and jointing techniques used are as accurate as possible.

For single and multifibre splices a number of simple bonding and fusion-splicing techniques are currently available and ruggedised or permanent multifibre connectors are frequently used to connect the fibre optic cables to distribution frames and optoelectronic equipment.

Mechanical splicing

Mechanical splicers join two fibres together by either clamping them within a structure or gluing them together with some form of index-matching adhesive. Losses are much higher than those found in fusion splices but the main advantage of mechanical splicing is the simplicity of operation and the tools required.

V-groove splicer

One of the simplest mechanical splicers is the **v-groove splicer** where the fibres are merely confined between two self-aligning v-groove plates. Some precision v-groove splicers employ **hold-down springs** made of berylium copper and lever activated rollers for dual-axis control. The principle of the v-groove splicer is illustrated in Figure 6.9.

A rapid curing index-matching adhesive is applied to the fibre ends to make a permanent fixture. The splice connector is then crimped onto the two buffer tubes. To ensure that attenuation caused by the joint does not increase due to seasonal temperature changes, it is essential that the adhesive used is age-resistant and an exact index-match of the fibre.

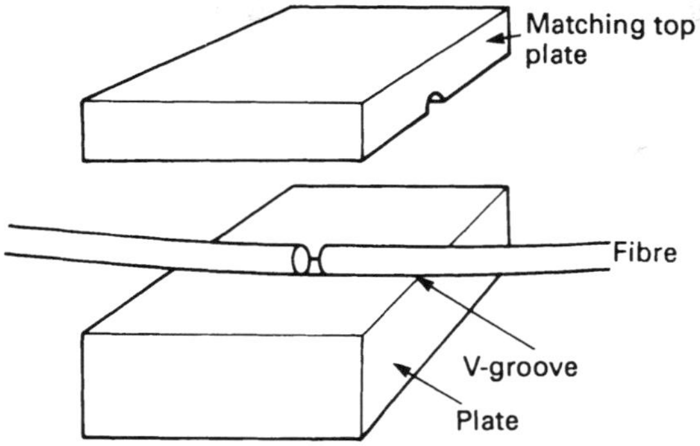

Figure 6.9 V-groove mechanical splicer, for joining optical fibres

Provided that extreme care is taken when completing the splice, an accurate cleave is achieved, and the right adhesive used etc, splice losses of less than 0.1 dB are possible.

Elastomeric splicer

Another type of mechanical splicer is the **elastomeric splicer**, whose principle is illustrated in Figure 6.10. It is similar to the v-groove splicer, but the plates are made of a flexible plastic material. An index-matching adhesive is then used to fuse the two fibre ends together or, alternatively, the splice connector can be crimped to the optical fibre cable.

Figure 6.10 Principle of an elastomeric splicer, which aligns fibres in a hole in the flexible plate

As these splicing methods are purely mechanical and there are no naked flames, these types of splices are particularly useful in hazardous areas.

The main disadvantage associated with mechanical splices is the possible loss introduced by an imperfect butt joint. These types of joints are therefore only suitable for short term, short distance, or temporary joints.

Mechanical field splicing kits

Low-loss mechanical splices are, however, possible using one of the ultra-violet splicers that are available on the market today. The principle of a typical ultra-violet splicer is illustrated in Figure 6.11.

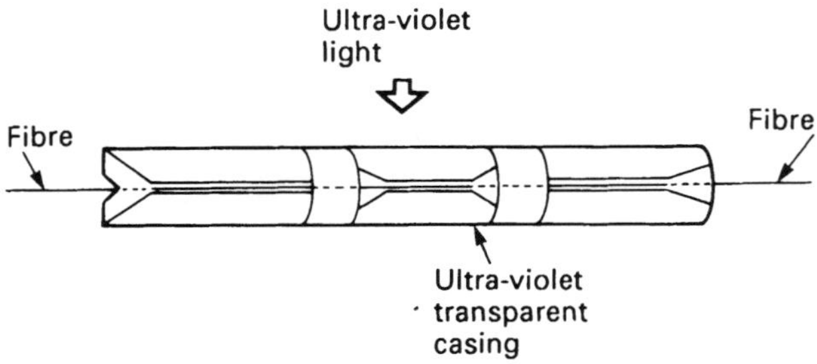

Figure 6.11 Ultra-violet elastomeric splicing principle

The procedure is very simple and, following fibre preparation (ie, precision cleaving, removal of a small portion of the cable sheath or buffer tube and thoroughly cleaning the exposed fibre), one or two drops of ultra-violet-cured adhesive are applied to the end of one fibre before it is inserted into one end of the splicer.

The end of the other fibre is similarly treated and is then pushed into the other end of the splicer until it butts against the first fibre. The splice is then cured by the application of an ultra-violet light source: splice losses of less than 0.2 dBs are not uncommon.

Fusion splicing

Fusion splicing requires the alignment of the two ends with microscopic precision and the precisely controlled application of energy to fuse the two ends together. Through automation, manufacturers have been able to make the task simpler and there are many different types of fusion splicers available on the market today although, because of the accuracy required, they are still very expensive; sometimes costing thousands of pounds.

Virtually all fusion splicers nowadays use **electric arcs**

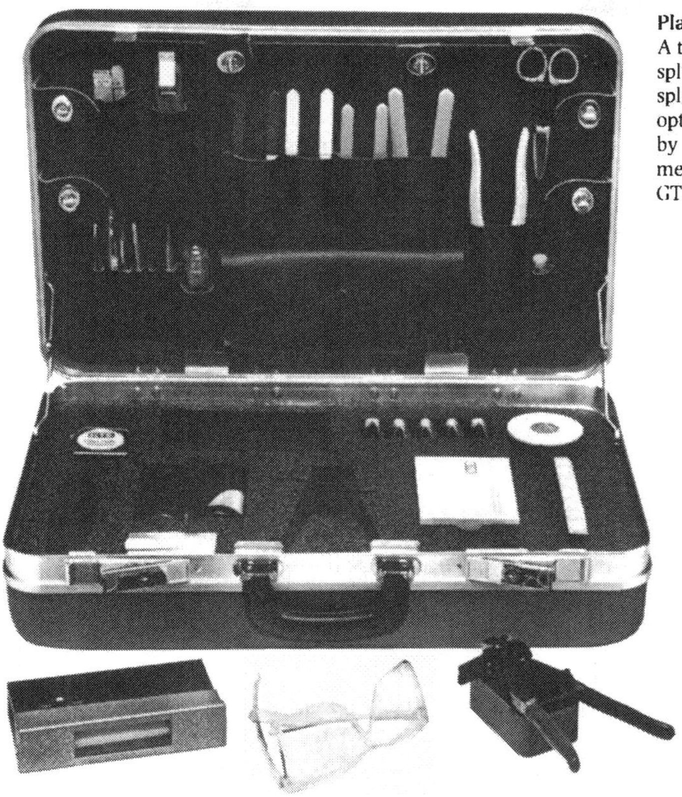

Plate 6.1
A typical field splicing kit, for splicing together optical fibre ends by ultra-violet means (courtesy GTE)

and have a **projection microscope**, (generally a binocular model with a 50 × magnification factor or more to enable the operator to observe the fibres during alignment), a method of comparing the amount of optical power transmitted through the fibre before and after splicing and a cleaver to cut the fibre to a precise perpendicular surface.

Quite frequently the equipment will also have a facility for coating the fibre ends after stripping.

Fusion splicing of graded-index fibres

This is the preferred method for long term junctions. The fibres to be joined are first stripped of their protective plastic coating. The ends are then precisely cleaved to produce end faces that are, within three degrees, perpendicular to

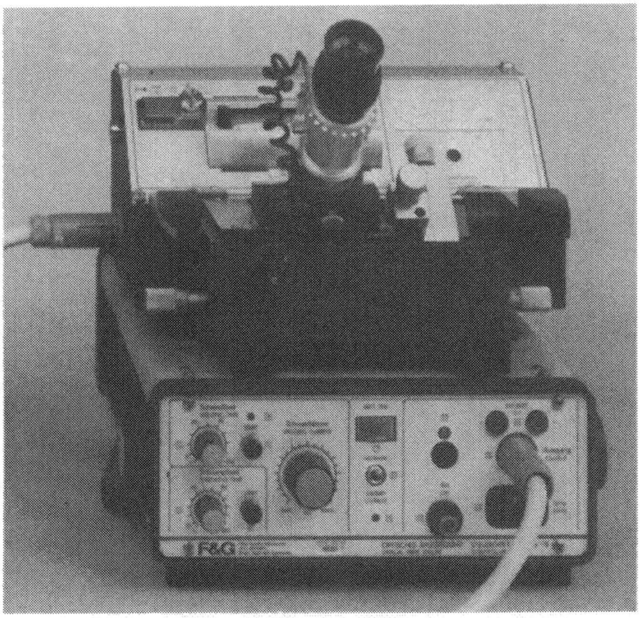

Plate 6.2
A typical fibre fusion splicing unit. The Philips OS615 fibre splicing unit has an arc-welding device to fuse the fibres together, and a heat-shrinking mechanism to allow heat-shrink sleeving to be applied after the weld to protect the joint (courtesy Philips)

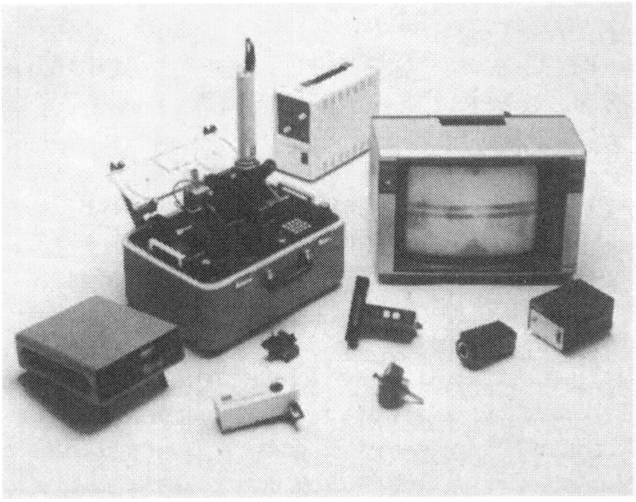

Plate 6.3
A typical fibre fusion splicing unit. The BICC AFS 3100 quick-splicing kit is suitable for reliable connections of 120 µm multimode fibres, with a resultant loss of less than 0·1 dB (courtesy BICC)

the fibre axis. The fibres are then placed in guides where they are automatically aligned and pushed together tightly enough so as to produce a visible compression when viewed through the microscope. The arc is then fired to weld the two fibre ends together. After the joint has cooled, it can be recoated with a plastic material to protect it against environmental degradation and make the outer diameter close to that of the original coated fibre.

Further protection can be given by crimping a splice connector over the two buffer tubes and pouring an air-cured **silicone rubber** over the splice connector to recoat the joint.

Fusion splicing of single-mode fibres

Although similar splicing techniques *can* be used for single-mode fibres, the precision and accuracy required for cutting and splicing two such fibres together (perhaps with a core diameter of 10 μm or less) is far greater than that previously used. Consequently, single-mode optical fibre fusion splices are not common.

Enclosures and containers

Copper cables use a variety of enclosures and containers in which to protect their joints, terminations and other components at the joining or branching point. Typically, these may be in a manhole, on a utility pole, or in special boxes above or below ground level.

One of the most important changes brought about by the use of optical fibres is that the size of the joint enclosure has had to be increased, to allow sufficient room to neatly arrange the splices and to provide enough space to ensure the bending diameter of the cable is within permissible limits.

Fibre splice enclosures therefore have to be designed so as to hold cable strength members tightly together, to block entrance of water, to retain gas (if the cable is pressurised), to electrically bond and ground any metal elements in the cable (such as strength members and

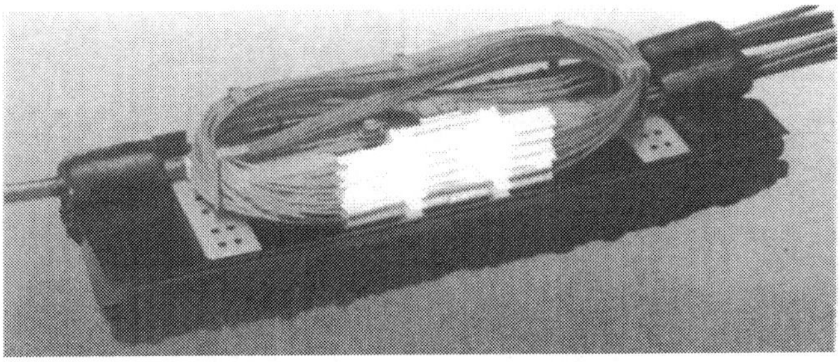

Plate 6.4
The Siemens universal closure 4-6, comprising a closure body of a highly stable polypropylene co-polymer, a sealing system with a corrosion-resistant sealing compound, and an inner metallic frame for mechanical cable sheath jointing and electrical connections (courtesy Siemens)

armour) and be capable of being re-entered, if required, to facilitate splice changes or repairs. Occasionally, extra space is also required for loading coils, jointing of cable shields, sheaths and strengthening elements.

Fibre splice enclosures should also be capable of allowing organisation of splices and fibres so they can be easily recognised and providing room for initial splicing and future modifications.

Fibres in multifibre optical cables are typically arranged in groups or units inside an enclosure. Normally each unit contains, about, 10 individual fibre splices and the required number of units are arranged inside an enclosure made for that particular cable.

Currently, closures are readily available on the market to cater for all needs whether it be for 2 or 2000 fibre splices.

Repeaters and regenerators

Containers are also manufactured for optical fibre repeaters and regenerators. These are normally made of steel which have been hot-galvanised inside and out, and incorporate a plastic-based multilayer anti-corrosion coating. This coating enables the container to be buried in the soil where it will be subject to only slight external temperature variations and is extremely difficult to access by unauthorised personnel [27].

The prime function of a repeater or regenerator is to receive the optical signal, convert it into an electrical signal, amplify (or regenerate) the information and then

Plate 6.5
The Siemens universal closure 6-18, which can accommodate up to 120 individual fibres in 12 units (courtesy Siemens)

convert it back into an optical signal. A repeater or regenerator is, therefore, required whenever the optical signal loss reaches a critical level. This depends upon a number of factors such as transmitter source type, transmitter wavelength, type of fibre, and transmission rate.

The use of optical fibres compared with copper cables reduces the numbers of repeaters (in the case of *analog* signal transmission) and regenerators (in the case of *digital* transmission) that are required. Typically, a 100 Mbit s^{-1} copper coaxial link, because of the inherent attenuation introduced by the cable, requires regenerators every 2 km or so. Using laser diodes and graded-index fibre for the same circuits on the other hand, would mean that regeneration would only be required approximately every 20 km. Looking towards the future, modern monomode fibres are now being produced with 200 GHz bandwidths – when in operation these will only need regenerating about every 100 km [8].

From Table 6.1 (page 106) the following conclusions about regenerator spacings may be drawn:

- graded-index fibre, for a wavelength of 850 nm, is well suited for 2, 8 and possibly 34 Mbit s^{-1} applications. If LEDs are used, regenerator spacings would have to be approximately every 8–13 km, depending upon the transmission rate. When laser diodes are used, attenuation as opposed to bandwidth, tends to be the limiting factor and regenerator spacings are increased to 10–16 km, depending on transmission rate

Table 6.1 *Showing typical regenerator spacings required for graded-index multimode and single-mode fibres, at various transmission rates and wavelengths*

		LED		*Laser*	
Regenerator spacing limited by:		*Attenuation*	*Bandwidth*	*Attenuation*	*Bandwidth*
Graded-index fibre for 850 nm (400 MHz km)					
Transmission rate	2 Mbit s^{-1}	8–12 km		12–16 km	
	8 Mbit s^{-1}		11 km	10–15 km	
	34 Mbit s^{-1}	5 km	10–13 km		
Graded-index fibre for 1300 nm (1300 MHz km)					
Transmission rate	34 Mbit s^{-1}		12 km	17–30 km	
	140 Mbit s^{-1}		7 km		20 km
Single-mode fibre for 1300 nm (20 GHz km)					
Transmission rate	140 Mbit s^{-1}			20–40 km	
	565 Mbit s^{-1}			16–30 km	

- graded-index fibre is also suitable for a wavelength of 1300 nm, together with 34 or 140 Mbit s^{-1} transmission rates. Use of LEDs is limited by the amount of material dispersion but distances can be increased by using laser diodes. The regenerator spacing in 34 Mbit s^{-1} systems is dependent upon the link loss, while at 140 Mbit s^{-1} it is limited by modal distortion. Any increase above 140 Mbit s^{-1} is not feasible due to the limited bandwidth
- single-mode fibres, at a wavelength of 1300 nm, are suitable for high bit rates and large section lengths, repeater spacing being limited almost entirely by link loss.

Although current technology tends to limit the upper transmission rate to 564 Mbit s^{-1}, with the use of VLSI (very large scale integration), transmission rates greater than 1 Gbit s^{-1} are becoming more common.

Laser amplifier repeater

In 1986 an experimental laser amplifier repeater was successfully demonstrated by British Telecom on a

120 km fibre link connecting two telephone exchanges in south-east England [28].

The amplifier was based on a semiconductor lens that received the incoming pulse of laser light at one end and emitted an amplified pulse at the other. Effectively this is an optical equivalent of the **travelling wave tube** used to amplify microwave radio. Demonstrations have shown that the amplifier is capable of simultaneously amplifying several separate sets of pulses at different light frequencies using the technique of wavelength division multiplexing.

The particular amplifier used by British Telecom was a **double channel planer buried heterostructure laser**, which was 500 μm long and whose laser faces had been given an anti-reflective coating to reduce the reflectivity of the laser face over 500 times (ie from a typical value of 40 per cent to 0.08 per cent) and the amplifier gain was 17 dB. The transmitter used was a **distributed feedback laser** emitting at 1.5 μm with a launched power of −2 dB. The receiver was a **PIN-FET** module.

It is anticipated that owing to their high gain, low power consumption and single-mode waveguide structure, semiconductor lasers will play an important role in future optical systems.

Multiport couplers

Optical couplers having more than one optical port to couple light in and out are called **multiport** devices. There are two main types of multiport couplers, those that provide three or four ports and those that can connect as many as 16 subscribers. These are called **mixers** and they depend upon the light being laterally transferred from one waveguide to another through the cladding. The amount of light being coupled depends upon the amount of overlap that exists at the fibre junction.

It should be noted that extreme care has to be taken when measuring the output power of a multiport coupler so as to ensure that the termination of the individual ports is reflection free. This is so that the emerging wave does not produce another wave, which re-enters the coupler.

Various examples, of multiport couplers exist, including:

- **symmetrical coupler**, or **switch** (Figure 6.12)
- **symmetrical offset coupler** (Figure 6.13)
- **asymmetrical offset coupler** (Figure 6.14); in this context the word *asymmetrical* implies that at least one section of the coupling is of a different fibre type. Asymmetrical offset couplers can be manufactured with low insertion loss and are suited to step-index

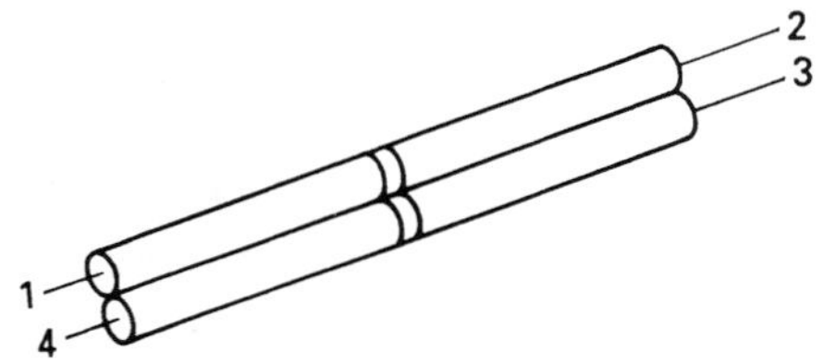

Figure 6.12 Symmetrical coupler, or switch

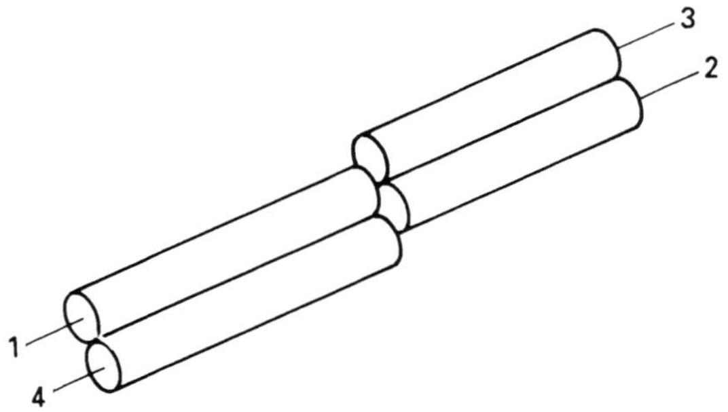

Figure 6.13 Symmetrical offset coupler

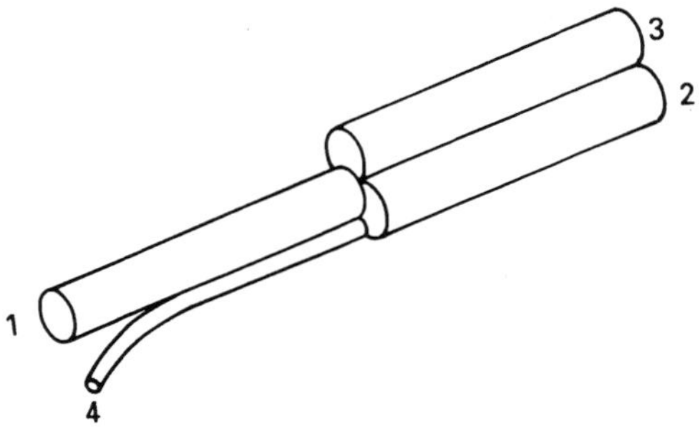

Figure 6.14 Asymmetrical offset coupler

fibre systems provided that the fibre cladding is thin. They are not suited to graded-index fibre because of the possibility of a mismatch being caused by the offset positions

- **asymmetrical three-port coupler with lapped core** (Figure 6.15); the lapped core of fibre 3 enables precise alignment in the axis of fibres 2 and 3, and hence low transmission loss for path 2–3 as well as 1–2
- **four-port coupler with beam-splitting mirror** (Figure 6.16); precise alignment of the fibre section is assured by use of the beam-splitting mirror. This coupler has no mode dependence (apart from polarisation effects) and can be used with step- or graded-index fibres. Through choice of mirrors, dielectric and laser design, light division can be varied and the multilayer acts as a wavelength selective filter

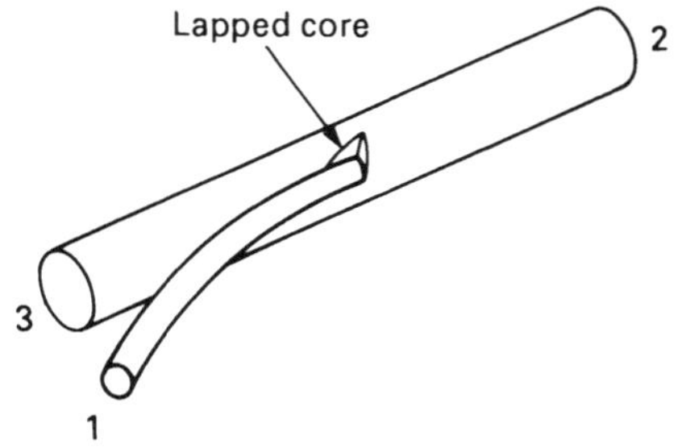

Figure 6.15 Asymmetrical three-port coupler, with lapped core

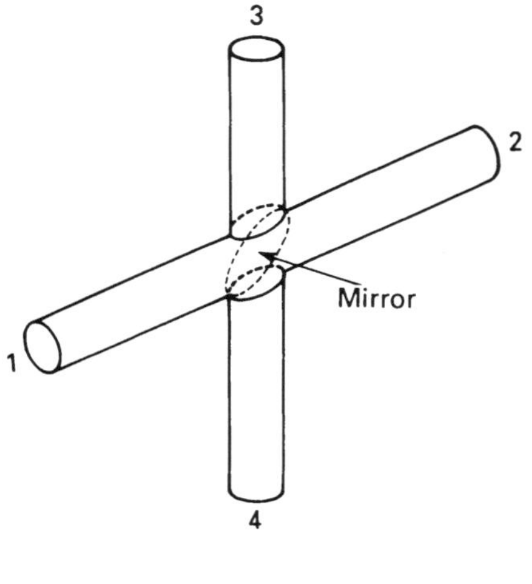

Figure 6.16 Four-port with beam-splitting mirror

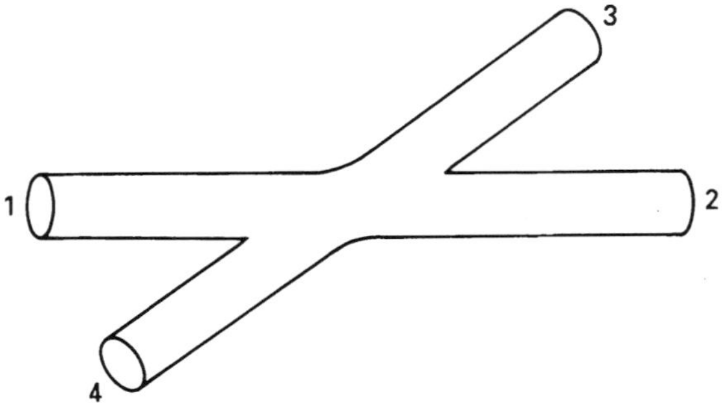

Figure 6.17
Four-port coupler (bitaper)

- **four-port coupler (bitaper)** (Figure 6.17); in the fused section, the fibre cores are separated from each other but the core modes are converted to cladding modes which provide optical power coupling from one fibre to another
- **couplers as switches** (Figure 6.18); because of their wavelength independence, couplers can also act as switches. These can range from a simple form (eg, the symmetrical coupler of Figure 6.12), to an electromagnetically operated four port switch, such as that shown in Figure 6.18. Principle of operation is very simple. When the reed (6) is energised, the four fibres are coupled together

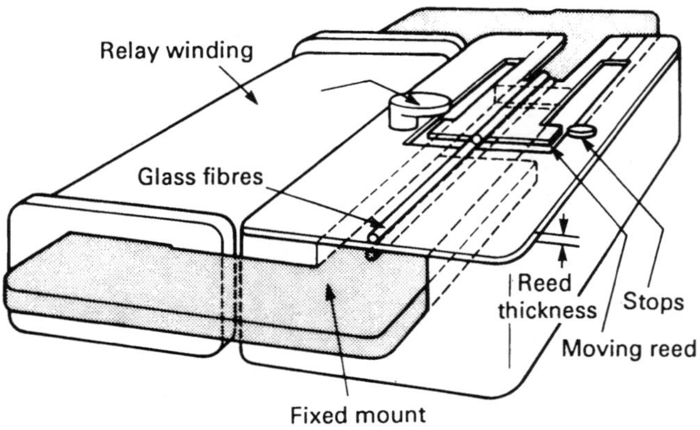

Figure 6.18
Electro-magnetically operated four-port switch

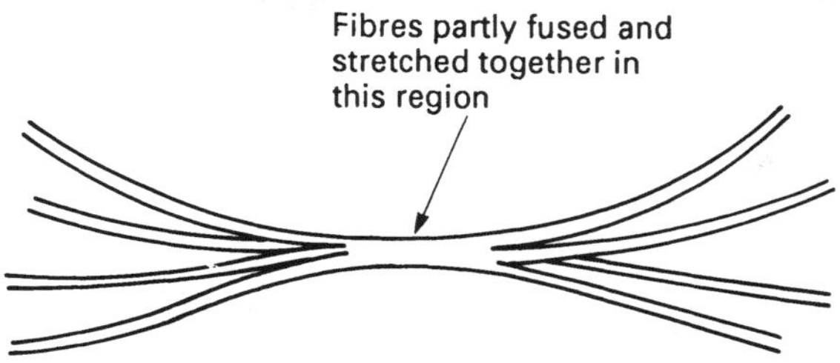

Figure 6.19 Fused-fibre, transmissive-mode, directional star coupler

- **transmissive and reflective star couplers** (Figure 6.19 and 6.20); **transmissive mixers** are made by biconically tapering and fusing together a number of fibres. The resultant changes in the fibre cause the incoming light to strike the side of the fibre cladding at a steep angle which, as it exceeds the acceptance angle, transfers some of the energy into the cladding. At the point where the cross-sectional area widens, the light returns the core region of the other fibres. Fibres enter a **reflective mixer** at one end while light is reflected off a mirrored surface at the other end. In the examples shown in Figures 6.19 and 6.20 the power from all the input ports is spread amongst all the output ports, to enable multiterminal links to be used over a shared fibre network.

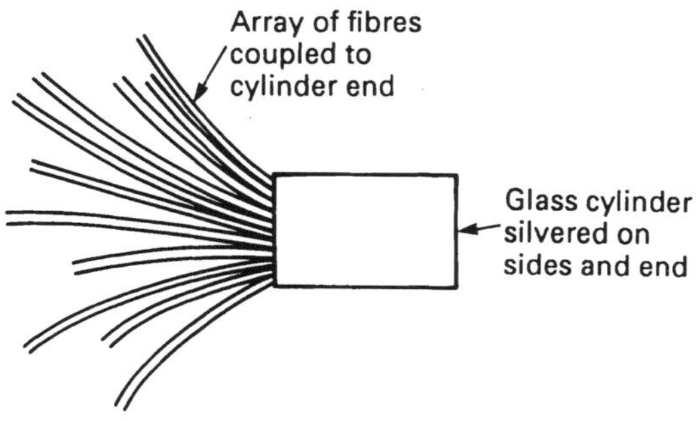

Figure 6.20 Reflective-mode, directional star coupler

7
Communication systems

Telephone networks are divided into two distinct levels, local and long distance. In **local networks** the subscriber is connected to the local or **terminating exchange** with distribution and main cables. These are known collectively as local **branch cables**. Local exchanges are in turn, interconnected with local **inter-exchange trunks.**

In long distance networks there are again two levels, regional and national. The **regional network** begins at the terminating exchange and extends via the **nodal exchange** to the **main exchange** using **trunk lines**.

Long distance networks

Until the wider usage of optical fibres, both coaxial cables and balanced copper conductors were used in national long distance networks. Depending upon the method of transmission, these cables were designed either for FDM (for analog), or PCM (for digital) transmission.

With the increased use of digital transmission techniques and the abundance of information services, expansion of long-haul networks has become very necessary and nowadays, because of the loss, increased bandwidth and size etc of copper conductors, optical fibres are being used more and more.

Obviously it would not be cost effective to replace *all* of the existing copper cable networks, particularly if they were designed to cover a high-density local environment where subscribers were dropped off in small numbers but quite often.

Instead, optical fibres are far more suited to carrying multigroup (eg, 300) signal groups which in optoelectronic terminology are referred to as **highways**. Current technology suggests that the cost effectiveness of an optical fibre highway link decreases after about 10 km, and the use of some other method for very long distances has to be considered – for example, a **microwave link**.

Typical cable capacities currently available are:

- trunk lines – 2000 fibres or more
- branch lines – 100–200 fibres
- subscriber lines – 1–10 fibres.

By far the most expensive part of laying lengths of communication cable underground is not the actual manufacturing costs but the **excavation**, **trenching** and **refurbishment** of cable ducts. Previously whenever a new underground cable was being installed, the possibility of including other cables that were routed in the same direction and capable of being laid in the same trench had to be considered. In some cases this meant that one trench was sufficient for all the current requirements and quite frequently this provided cost sharing advantages.

Optical fibres, because of their size, lend themselves to sharing not only one trench but in some cases, the same cable sheath as other fibres, coaxial, or copper quad cables.

An alternative method of cable distribution across non-populated areas (other than direct burying) is to utilise existing high voltage overhead power line pylons to support the cable, as illustrated in Figure 7.1.

In this method the non-metallic optical fibre cable is incorporated into the phase conductors or ground wires. Because of its low weight, the optical fibre cable subjects minimal additional stress onto the pylon.

There are, of course, disadvantages to using optical fibre cables that have been trenched or ducted underground – the most obvious being the damage that rodents can do to the cable. This problem can be overcome by using steel tape over the cable, but this is very expensive and is also against the basic principle of providing a completely non-metallic cable. Another method of

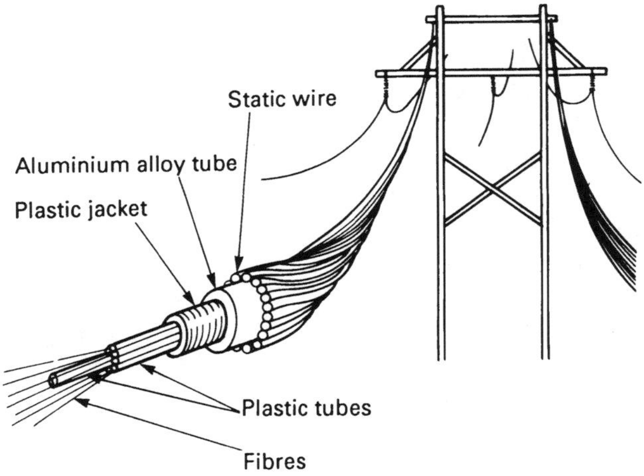

Figure 7.1 Ground wire, incorporating optical fibres, mounted on the top of a pylon power line structure

overcoming the problems associated with rodents is to encase the cable in **resin-impregnated glass** or **polyamides**, but this too, increases the cost of the cable.

Already a great deal has been learnt from previous long-haul optical fibre installations particularly in the national and international fields:

- **Deutsche Bundespost**: on 1 September 1977, almost exactly 100 years after the Deutsche Bundespost opened its first public telephone circuit, the first optical fibre cable was installed into a telephone network, in Germany. This was an optical link incorporating a 34 Mbit s^{-1} digital transmission system [9]
- **Alpine rescue huts**: in the past, emergency radios and radio telephones located in Alpine rescue huts were frequently out of action because of power problems, lightning strikes and cables breaking because of falling trees and radial ice etc. Optical fibres have now successfully replaced the previous fault-prone copper cables and battery problems have been overcome by using self-contained solar power plants at each hut
- **British Telecom**: in 1983, STC was contracted by British Telecom to install the first of a number of optical fibre links which were laid between Luton and Milton

Keynes, (a total distance of 28 km) following the route of the original Roman road.

If coaxial cable had been used for this route at least 14 regenerators would have had to have been used. As it was, by using far more efficient and reliable optical fibre cable, *no* regenerators were required. The optical fibre was laid in 1.5 km prejointed sections without the use of any specialised equipment. Extreme care was taken at section joints to ensure that the splices did not introduce additional attenuation. This was achieved by fusion (arc) splicing (see Chapter 6).

The UK network is already well advanced in the use of digital and optical technology. All 53 trunk units are now digital, over 1000 of the 6000 local exchanges (central-offices) are wholly or partly digital and it is anticipated that trunk transmission systems will all be digital by the end of 1989 [29].

By January 1988 British Telecom had over half a million fibre kilometres in use as follows:

- trunk – 233 000 fibre km
- junction – 173 fibre km
- local 31 000 fibre km
- **Submarine cable**: successful underwater trials have been completed at Loch Fyre in Scotland by STC to prove the feasibility of using optical fibre cables for long-haul submarine links and links are being planned between Portsmouth and the Isle of Wight, UK and Belgium and in the not-too-distant future transatlantic routes, where it is anticipated that an underwater optical fibre life span of 30 years will be possible.

  PTAT Systems and Cable & Wireless (Mercury) have also joined forces in owning and operating a $350 million transatlantic optical fibre cable linking North America and Europe. PTAT-1 (which stands for Private Transatlantic Telecommunications System – phase 1) will begin operation in mid-1989 and will consist of three operation fibre pairs, each carrying three 140 Mbit s^{-1} streams, capable of carrying 18 000 simultaneous telephone calls. The cable will have a length of 7000 nautical miles and the system will include 114 repeaters.

PTAT-1 will be followed, sometime in 1992 by PTAT-2 and these will become part of Cable & Wireless's *Global Digital Highway* which, as the name implies will be a world-wide optical fibre system linking all continents.

STC Submarine Systems of London have recently been awarded a contract to develop an 84 mile undersea optical fibre cable connecting Dartmouth and Jersey in the Channel Isles. This single-mode system will operate at data speeds up to 140 Kbit s^{-1} and will consist of six fibre pairs. The system is due to begin service in the Spring of 1989 and is reported to be the world's longest link without repeaters [18].

Local networks

It is interesting to note that there are seven colours in the rainbow, but until the 1960s the *only* colour available for telephones was black! Then the *natural monopoly philosophy*, which effectively dictated the type of service that could be made available to subscribers – as opposed to providing a service that fulfilled the customers requirements – came to an end. Since that time we have moved from two 'phones on the desk (one for external, the other for internal calls) and gone through four generations of switching systems, from analog to digital, from copper to optical fibre networks and from the *system is the solution* to what the customers needs. Figure 7.2 shows how optical fibre systems can replace existing copper local networks.

Local cable systems (for example **cable television**) are today experiencing extreme difficulty in selling services over existing copper cable networks and only a minority of homes having access to these facilities actually subscribe to them [30]. Even such systems as the French

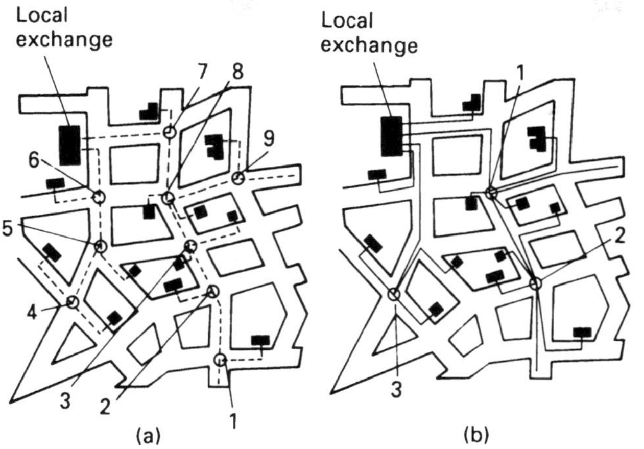

Figure 7.2
Optical fibre local networks can replace existing copper cable local networks to the advantage of the network provider (a) a copper cable network requires more repeaters or regenerators, and so a large number of closurers is required, ten is this example (b) optical fibre local networks require fewer regenerators, so fewer closures are needed, only three in this example

Minitel system whereby subscribers are provided with a terminal instead of a telephone book and the facility of accessing weather reports, stock quotes and other specialised services over a copper network that joins together approximately 3.5 million users, is not fully utilised. Although attracting a great deal of interest it still lacks the ability to provide today's subscriber with *exactly* what he requires ie, an infinite amount of information, available at the flick of a switch or the press of a button.

This situation is of course not being overlooked by the cable network companies who are fully aware of the growing phenomenon of **telecommuters** and the accepted fact that by the year 2000, one in three workers will be operating out of his or her home. It is for precisely this reason that more and more interest is now being shown in the optical fibre market.

In spite of the increased cost and difficult jointing procedures, the many advantages associated with optical fibres over normal copper and coaxial cables ensures that fibres will be used more and more to provide **switched, broadband communication services** and networks for both business (**videotelephony**, **videoconferencing** and **videotex**) and private (videotelephony, **multichannel radio/television programming** and **armchair shopping**) subscribers.

Distribution of stereo sound programs and television programs are conventionally provided by sound and TV broadcasting services. In an optical, broadband **local area network**, the users are arranged in a star configuration and the subscriber uses individual **program selection channels** to call up the desired program from the **network control office**. One of the advantages of this approach over the conventional distribution technique is that the number of distributable programs is basically unlimited. To ensure that each subscriber can be accessed at a subscriber outlet by several people simultaneously, there must, of course, be sufficient transmission channels available for each service. When planning local area subscriber networks as well as providing for the current requirement, future expansion should also be anticipated. In copper cable systems this would have meant the inclusion of additional cable pairs. In optical fibre systems the availability of transmission modes virtually guarantees that sufficient channels will *always* be available. The layout of subscriber networks quite naturally varies from area to area and country to country, but typically it would be made up of about 16 narrowband 16 Kbit s^{-1} channels for telephone or data transmission, two 16 Kbit s^{-1} channels available for **subscriber video communication equipment**, four stereo sound channels, four broadband channels for stereo TV and eight or so program selection and monitoring channels.

At the subscriber premises the line will be terminated in a **subscriber line terminating unit** which converts the optical signals into an electrical format and vice versa. The unit also performs some of the **multiplexing** and **demultiplexing** of the signals transmitted via the indoor cable between the line unit and associated equipment terminal. At the **central switch**, a **line termination module** acts as a **muldex** for the **switched services** being supplied to the subscriber and several signals, separated by **time division multiplex**, are incorporated with a **binary line signal** and digitally transmitted over the optical wave-

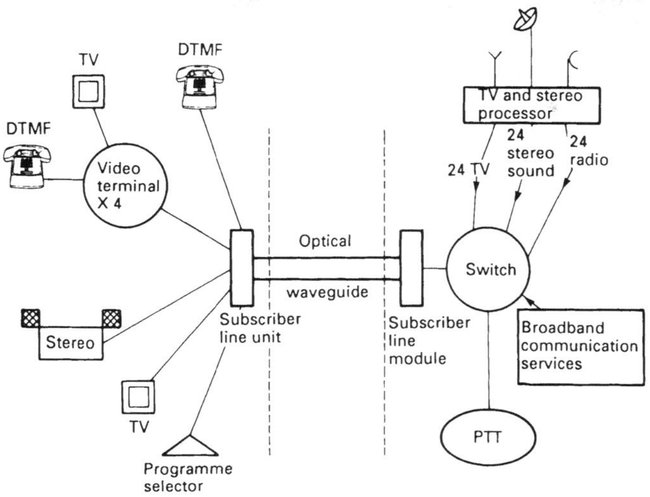

Figure 7.3 Subscriber communication services system, using an optical fibre waveguide

guide to the subscriber. Such a system is illustrated in Figure 7.3. Mainly for economic and operational reasons, subscriber lines are, as a general rule, operated on an unrepeated basis. Although the cost of producing optical fibre cable is still fairly high, it is still more cost effective to install an optical system as opposed to a coaxial one, as can be seen from Figure 7.4.

Previous experience with local networks is an essential consideration when planning new systems. For instance, some of the data obtained by Siemens engineers from one of the company's North American projects [31] where approximately 4000 km of fibres were installed in a cable route of about 70 km, have produced some interesting results. Two of the main lessons learned were with regard to the **tensile characteristics**, the cables showed no significant increase in attenuation ($<0.1\,\mathrm{dB\,km^{-1}}$ at a wavelength of 850 nm) in response to a tensile force of 2500 N and the reaction to temperature changes (in the temperature range from -30 to $+60$°C) was that the attenuation variations were always less than $0.8\,\mathrm{dB\,km^{-1}}$ at a wavelength of 850 nm. There are many other examples such as:

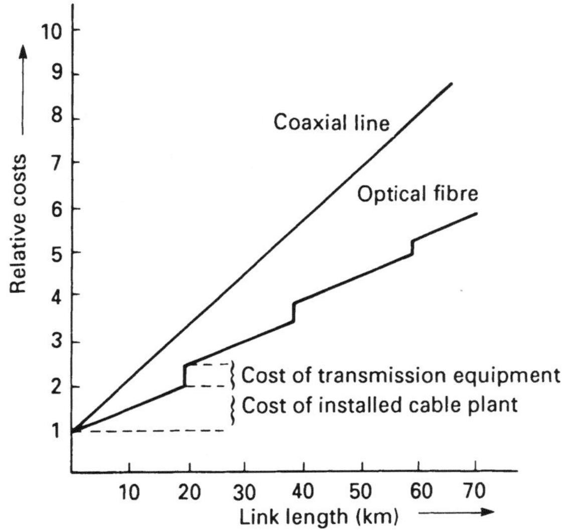

Figure 7.4 Cost comparison between coaxial and optical fibre systems for a transmission rate of 565 Mbit s^{-1}

- **World Trade Centre (New York City)**: the system installed in the World Trade Centre in New York is an example of an in-built optical fibre system. In August and September 1981, six reels of six-fibre cable were installed and connected vertically from floor B1 to the 7th, 41st, 74th, 75th, 108th and 110th floors. Installation of the cable took about a week. All the cables were pulled by hand into a common 100 mm duct which also housed a number of conventional copper cables. Pull-in boxes were provided approximately every fifth floor for cable access [32].
- **Deutsche Bundespost BIGFON project**: normal subscriber loop networks require an individual copper pair for each telephone service and separate high quality, low-loss coaxial cables for TV radio and videotelephony services.

  In the **BIGFON** project (an acronym for *broadband integrated glass fibre local communication network*), the Deutsche Bundespost set up a number of trial subscriber networks in various German cities using optical fibre cable. In the future it is intended that these will be interconnected for long-haul applications.

Each subscriber is provided with 16 channels at 64 Kbit s^{-1} which enables use of normal telecommunication services such as telephony, facsimile, teletext, videotext etc. In addition, the subscriber is also able to select up to four TV and radio channels from a virtually unlimited number of programs available from the local switching centre. With the addition of a video camera and a television set, each subscriber can form a videotelephony link. Because of the high transmission bandwidth, low attenuation of graded-index fibres, and by using wavelength division multiplex, only one optical fibre is required in each direction (ie, a total of two fibres per subscriber). As there is still a limited number of subscribers and because no standards for TV video signal and digital picture signal have as yet been internationally agreed, BIGFON transmits its video telephony signals as frequency modulated signals. The long term objective is, of course, for the transmission of all signals to be in a digital form utilising an **integrated network**.

Compatibility between visual communication and TV program distribution is essential, to enable the same equipment to be used for either system and thus reduce overall costs. Future visual communication channels must be the same quality as a broadcast standard, with 625 lines per frame, in accordance with CCIR.

With the exception of optical fibre cable, the BIGFON network utilises basic communication principles and is very similar to other subscriber networks that use normal copper cable technology, except that the tree structure normally used in large cable television systems is unsuitable for a network with a subscriber-orientated, bidirectional narrow and wideband communication capability, especially since switching channels are not available. Instead, in an **integrated optical network** such as BIGFON, the subscribers are arranged in a star configuration around the local switching centre, which is then capable of providing individual communication services at various bit rates to suitable interfaces over twin-fibre cables. Normally multicore cables (10, 20 and 30 fibre bundles) are used to connect the subscribers to the local switching centre [33] [34].

Looking to the future, The Deutsche Bundespost has suggested that it has now become technically and economically feasible to consider the early installation of single-mode optical fibres into *all* commercial and domestic properties

- **British Telecom's London dealer-interlink optical fibre network**: the dealer-interlink optical fibre network is primarily designed for City of London customers requiring a high quality, fast service to other customers within the city. Dealer-interlink is part of British Telecom's investment in its **flexible access system** (FAS) which will be extended to all major cities by 1992. Subscribers rent groups or types of circuit which are connected to a central point and which can be interlinked to other subscribers.

  To date (1989) over 60 000 km of optical fibre cable have been installed, and one building even has as many as 15 000 single mode fibres terminating in it [35]. This has been achieved at an initial cost of approximately £30M ($54M) and over 100 customers are currently using the service. In a similar scheme Mercury (a subsidiary of Cable & Wireless) already has almost 600 fibre km laid in London and are also in the process of installing similar networks in Manchester, Edinburgh and Birmingham
- **Norlight**: moving with the times, in efforts to develop their own internal telecommunication systems and diversify into profitable ventures, more and more companies are joining together and producing common optical fibre based systems. For example, in the USA, five major electric companies have formed a partnership called Norlight which has developed its own

network that stretches 650 miles between Chicago and Minneapolis.

The principle is very simple. The optical fibre cable is first wrapped in aluminium-clad steel strands, then strung along the tops of the power line support structures belonging to Norlight members (also see Figure 7.1). One of the advantages of this method is that the aluminium-encased cable will also function as a static ground wire for the power system.

Although the first phase of the Norlight system is estimated to have cost $100 million, it must be appreciated that as each Norlight member currently spends $12–15 million annually on telecommunications, even in the medium term this represents considerable savings.

The world's first fibre optic transatlantic cable, capable of carrying 40 000 simultaneous telephone conversations, opened for business in December 1988. The cable, known as TAT-8, links Europe and North America and is the result of a three way collaboration between British Telecom, AT & T and France Telecom.

Considerations in designing an optical fibre cable system

The most important considerations when designing any optical fibre system are attenuation loss and usable bandwidth. Both of these are a function of the amount of input power that is required to produce a preset or desirable output power. The system loss is given as a function of power out against power in and is measured in decibels per kilometre of cable (eg, $10 \log[P_{out}/P_{in}]$ in $dB\,Km^{-1}$).

Consideration must also be given to system expansion and this should not only include additions to the system and the necessity to splice new cables onto existing cables, but also excavation, cable routine etc.

The design engineer of any local network, be it private or business, will also need to be aware of the following points:

- the number of subscribers
- the application (ie, indoor, ducted, buried, aerial or submarine)
- the transmission properties (multichannel television, videoconferencing etc)
- the climatic and geographical environment (temperature, tension, pressure, ice or windloading)
- the local environmental restraints (for instance the requirement to utilise a special non flammable or rodent-resistant, jacketed cable).

Attenuation introduced by fibre splices (typically 0.05 to 0.02 dB), cable distribution cabinet jumpering, together with in-line or branch splices must also be kept to an absolute minimum. Instead of using main distribution frames (MDFs), this can be achieved by either laying separate cables to branch networks (as opposed to branch-splicing a cable off the trunk network) or, distributing cables in the switchroom directly from the trunk cable (ie, fanning out). Attenuation can naturally be minimised at spliced-cable sections if correctly dimensioned loose tube cables are used.

Digital and analog system designs

Digital

The only characteristic affecting the span length for digital transmission is the permissible attenuation between the optical transmitter and the optical receiver and the average received optical power necessary to achieve a certain bit error rate.

Detailed consideration of the theoretical, achievable, receiver sensitivity shows that provided the system is not bandwidth-limited, binary optical transmission will give a receiver sensitivity several decibels higher than multi-level optical transmission [36].

For systems up to 140 $Mbit\,s^{-1}$, graded-index fibres are used. Above 140 $Mbit\,s^{-1}$, especially in long-haul com-

Table 7.1 Optical fibre system types compared, showing typical regenerator spacings and system factors

	Transmit power dBm	Receiver power dBm	Connector loss dB	Theoretical system loss dB	Maximum section loss dB	Fibre loss dB km^{-1}	Splice loss dB km^{-1}	Cable loss dB km^{-1}	Regenerator spacing km	Number or channels
8 Mbit s^{-1}										
840/880 nm graded-index LED	−16	−56	3	37	31	3	0.2	3.5	9	120
840/880 nm graded-index laser	−2	−56	3	51	45	3	0.2	3.5	13	120
1300 nm graded-index LED	−17	−49	3	29	23	1	0.2	1.5	15	120
1300 nm graded-index laser	−2	−49	3	44	38	1	0.2	1.5	25	120
34 Mbit s^{-1}										
840/880 nm graded-index laser	−3	−52	3	46	40	3	0.2	3.5	11	480
1300 nm graded-index LED	−19	−47	3	25	19	0.8	0.1	0.9	21	480
1300 nm graded-index laser	−3	−47	3	41	35	0.8	0.1	0.9	39	480
1300 nm single-mode laser	−4	−47	4	39	33	0.5	0.1	0.6	53	480
140 Mbit s^{-1}										
1300 nm graded-index LED	−20	−40	3	17	11	0.8	0.1	0.9	7	1920
1300 nm graded-index laser	−3	−40	3	34	28	0.8	0.1	0.9	31	1920
1300 nm single-mode laser	−4	−43	4	35	29	0.5	0.1	0.6	48	1920
565 Mbit s^{-1}										
Single-mode laser	−4	−35	4	27	21	0.5	0.1	0.6	35	7680

munications, single-mode fibres are used as they provide the bandwidth required for the regenerators. Because the **spectral bandwidth** must be low to minimise material dispersion at high bit rates, laser diodes must also be used.

Analog

Apart from bandwidth and attenuation, linearity of the optoelectronic transmission path is also important for analog transmission. Although linearity can be improved by reducing the power to the transmitter, a reduction in the modulation factor will reduce the already small system-to-noise ratio.

This could be overcome by using negative feedback circuits (but these introduce excessive attenuation), having non-linear pre-equalisation (but diode losses are increased), or restricting the span length (but then the signal-to-noise ratio is reduced with power increases).

An analog system is thus really only feasible if the quality of the system is modest.

Channel and cable capacity

For applications having several communication links between the same points, it is economical to use some form of multiplexing to combine communication channels onto a single fibre-pair before transmission. **Time division multiplexing**, whereby a number of sampled messages are used to modulate a pulsed carrier at different time intervals thereby ensuring that at any one instance only one carrier is being transmitted, is widely used.

Typical regenerator spacings, wavelengths and transmitter types are compared in Table 7.1, with regard to other system factors.

8
Optoelectronic test techniques

In this age of Information Technology and the requirement for high-capacity wideband communication systems, it is of little wonder that industry is turning more and more to optical fibres and it has become increasingly obvious that there is a requirement for a completely new range of **optical measuring instruments**.

While the basic test requirements associated with optical fibre systems are of course fairly simple, the range of instruments currently available is enormous and it is essential for today's engineer to have a clear knowledge of exactly *what* he is testing, exactly *why* he should test it, and exactly *how* he should test it.

Similar to other communication systems there are two distinctly separate areas that require testing – *production* and *field*.

Production testing

The whole purpose of production testing is to ensure that the user is provided with an end-product consistency. Manufacturing optical fibres is a precise science and in the same way as in any other volume manufacturing process, production testing is instrumental in ensuring a uniformity of manufacture and a means of verifying that the material specifications, within each phase of the manufacturing process, are to the required quality. Production testing usually involves the following:

- measuring the **refractive index profile**; prior to coating the fibre with an acrylic material
- measuring the **spectral response** after coating; attenuation versus frequency

- checking **fibre geometry**; including concentricity of core and thickness of coating
- **backscatter measurement**; to detect breaks and other defects in the end product.

In order to carry out such an exhaustive series of checks a variety of sophisticated measuring instruments, as well as highly trained staff, are required. Neither of these commodities come cheaply and, as a general rule, manufacturers typically estimate to commit at least 30 per cent of their total manufacturing budget to production testing. Production testing therefore represents a significant proportion of the final product cost.

Another factor that has to be considered is the importance of ensuring that the test equipment used with optical fibre systems is always of the highest accuracy. To provide this level of confidence it is, therefore, essential that these instruments are regularly calibrated against a reference standard to detect, correlate, adjust and determine variations in their accuracy. This requirement is usually very expensive but hopefully will become cheaper as the technology progresses.

Field testing

When considering the purchase of an instrument, an engineer must be sure to take account of:

- the permitted range of fibre connectors; are adaptors provided, are they easy to fit, will the equipment accept a bare fibre etc
- the probable requirement to recalibrate the test equipment following the changing of connectors or adaptors
- whether detectors are available for all nominal wavelengths (ie, 850 nm, 1300 nm and 1550 nm)
- whether the equipment is truly portable (ie, battery powered).

Other than system bandwidth, probably the most important characteristic that has to be considered when engineering optical fibre cable systems is that of the losses due to fibre splices, connectors, junctions and, of course, the actual fibre itself. Thus, as well as the normal range of multimeters, oscilloscopes, powermeters and other test equipment that a typical repair facility would possess, there is now the added requirement to carry an additional amount of test equipment specifically designed for light measurement. As one would imagine, these are mainly concerned with the availability of a light source and the means of monitoring it.

Laser light sources

Most of the so-called portable laser light sources available prior to 1988 were ruggedised water-tight pieces of equipment, weighing about 5 kg or so and usually powered by rechargeable sealed lead-acid batteries capable of providing more than 20 hours' operating life. Industry quickly realised, however, that there was a requirement for a *truly* portable, drop-proof, multi-wavelength, reasonably priced piece of equipment and, thankfully, there are now a number of extremely efficient light sources available on the market at the time of publication, suitable for field maintenance and installation.

Operating these equipments so that loss measurements can be taken is usually very simple and normally consists of connecting the light source to one end of the fibre link and measuring the amount of light intensity (and, therefore, the fibre system power loss) at the other end with the aid of some kind of powermeter.

As one may suspect there are, of course, a number of specific problems directly associated with fibres – especially with respect to mechanical shock, excessive bending and inadequate insulation – that are not normally found in conventional copper cables. It has, therefore, been necessary to develop not only a completely new range of portable test instruments, but also – due to the range and complexity of some of the measurements

Plate 8.1
Philips OPS5 optical power source, together with the OPM6 optical powermeter. Semi-ruggedised, shock-protected field instruments. The OPS5 has two different power levels and, although calibrated specifically for three standard wavelengths (850, and 1350 and 1550 nm), will operate throughout the spectrum between 800 and 1800 nm (courtesy Philips)

that have to be taken – a completely new form of measuring technique.

Some of these instruments and measurements follow.

Power and attenuation measuring test sets

Power and attenuation measuring test sets are as fundamental to testing optical fibres as the simple continuity tester is to testing conventional copper wires and coaxial cables. They are invaluable during initial commissioning and in-service testing of an optoelectronic system.

Most power and attenuation test sets are normally very simple (merely consisting of an optical transmitter with an associated receiver) and there are many hand-held as well as bench-mounted models currently available from manufacturers such as Philips, HDW Elektronik, Wandel & Golterman, Hewlett Packard, Photodyne, Solatron, Thorn EMI etc. All of these manufacturers provide equipment with very similar facilities and in some respects their products often look similar.

Optical time domain reflectometer

Optical time domain reflectometry is a valuable and widely used measuring technique for optical fibre cable installations and was developed in the mid-1970s. The credit for the early development of optical time domain reflectometers (OTDRs) goes to Dr Stuart Persenick and Dr Michael Bainoski. The operation of an OTDR is based upon Rayleigh backscatttering, whereby a small amount of light is scattered in all directions as it travels down the fibre. An OTDR makes use of this phenomeon by transmitting pulses of light into a fibre and then measuring the reflections that occur. The principle is illustrated in Figure 8.1.

OTDRs are widely used for measuring the transmission performance and attenuation characteristics (ie, loss versus distance) of an optical fibre, especially when fibres are first installed or after they have undergone repair. An OTDR can also be used to detect and locate faults, splices and other discontinuities within fibres that are not normally visible to the naked eye.

There are many OTDRs on the market today but they are normally very expensive. Nevertheless the quest for cheaper alternatives continues.

Bandwidth measuring test sets

One of the limiting factors governing the transmission of information in an optical fibre is the operational bandwidth of the fibre itself. Although details of the operational bandwidth of the original fibre are always available from the manufacturer, in practice, when one or more fibres are spliced together in the field (owing to the possibility of less than perfect splices, junctions and mismatches etc) there is no real means of accurately predicting the resultant bandwidth.

In wideband applications this is, however, an exceed-

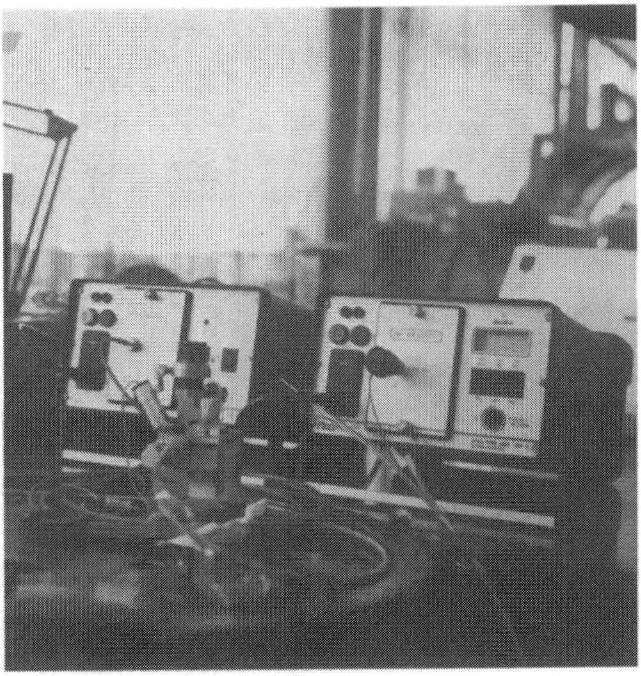

Plate 8.2
Philips OPS10 optical power source, together with the OPM10 powermeter. These are precision instruments. The OPS10 is a highly stable, temperature-controlled power source for LED or laser operation, while the OPM10 is capable of measuring absolute light levels in dBm, and attenuation in dB (courtesy Philips)

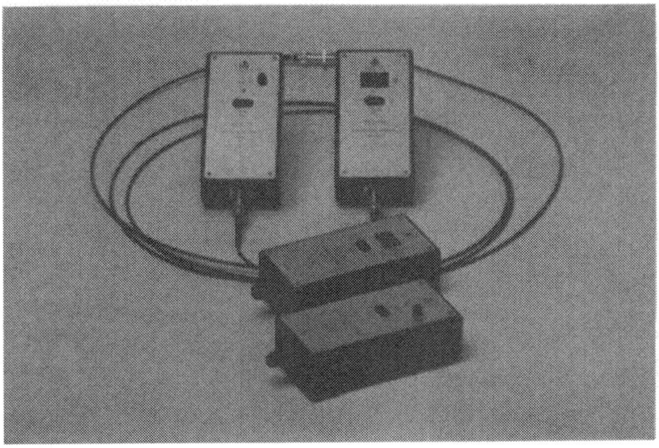

Plate 8.3
HDW Elektronik T53 test set. Available for both 850 and 1300 nm wavelengths, is not sensitive to external light and has an accuracy better than 0·5 dB (courtesy HDW Elektronik)

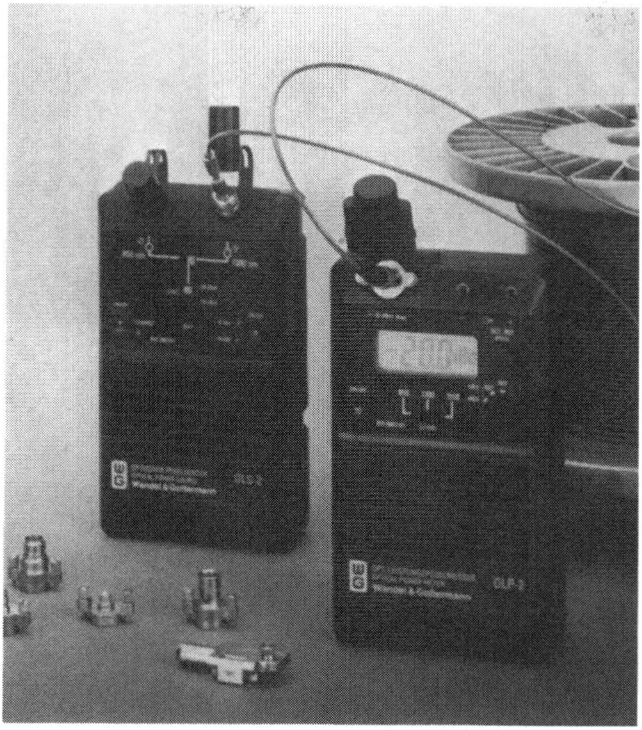

Plate 8.4
Wandel & Goltermann OLS2 light source, together with the OLP2 power level meter. The OLS2 features a special compensation circuit, ensuring the signal is highly stable, and when used with the OLP2 the technician has a lightweight (less than 1·2 kg total weight), compact, highly accurate test equipment set, covering all the normal wavelengths (courtesy Wandel & Goltermann)

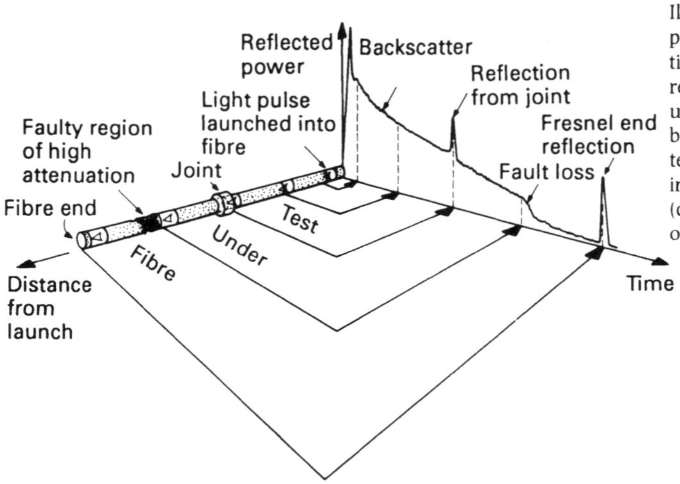

Figure 8.1
Illustrating the principle of optical time domain reflectometry, using Rayleigh backscattering to test optical fibre installations (diagram courtesy of STC)

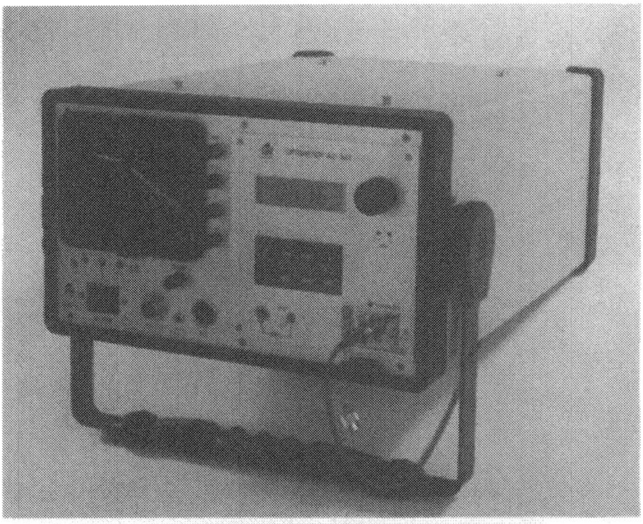

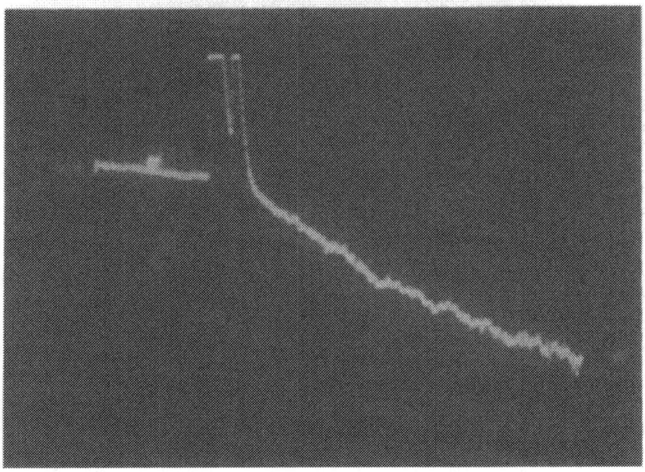

Plate 8.5
HDW Elektronik T52 optometer (above). This instrument is available for both 850 nm and 1300 nm wavelengths and has a dynamic range better than 65 dB. It features an XY plotter output and a CRT display obtained from the T52 optometer (below) (courtesy of HDW Elektronik)

ingly important aspect as it is essential for the installer, user and maintainer, to be aware of the precise bandwidth of the system after installation.

Once again, there are a number of equipments available to assist the engineer.

Stress measurement

As previously discussed, in order to achieve a long working life, it is essential that optical fibres are not exposed to

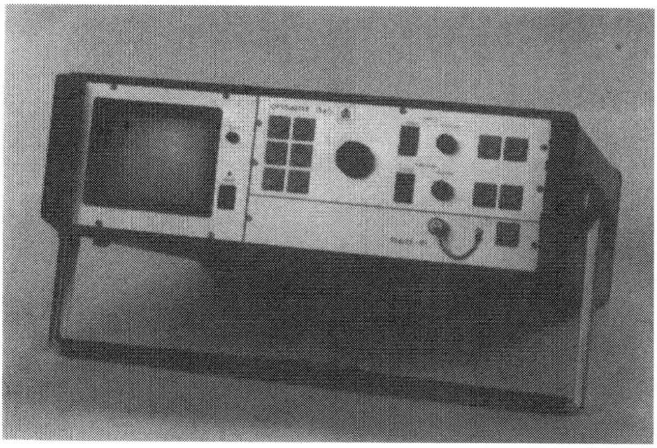

Plate 8.6
HDW Elektronik T54 optometer, catering for the 1300 nm and 1550 nm wavelengths. It features an XY plotter output (courtesy HDW Elektronik)

any undue mechanical stress or vast changes in temperature. There are, however, quite a number of situations during production and cable laying where some kind of stress may be unavoidable. In these cases it is necessary to have a means of measuring and determining the amount of stress that has, or is being, exerted onto that fibre. Measuring instruments are available for this.

Future developments

At the moment there are quite a number of different optical fibre connectors available for optoelectronic systems. This has been caused by different manufacturers having

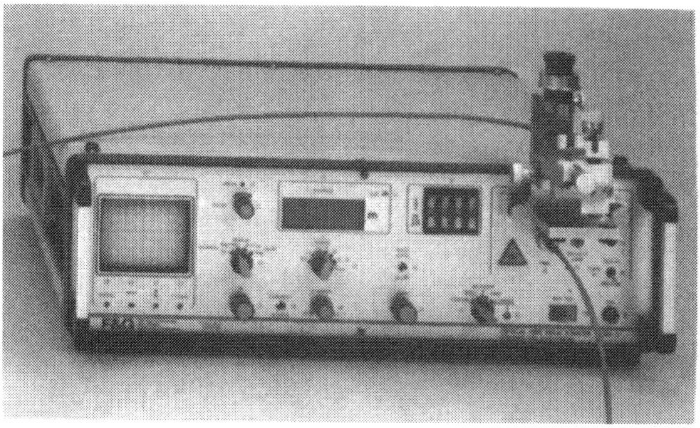

Plate 8.7
Philips OFR11 optical fibre reflectometer. The instrument covers all standard wavelengths and is suitable for quality during fibre manufacture, laboratory use, and fault location in the field. It is mains- or battery-powered, with an XY plotter output (courtesy Philips)

Table 8.1 *The variety of connector types available for use with optical fibre systems (information courtesy Wandel & Goltermann)*

Plug type	Similar type or alternative	Pin diameter	Main uses	Standard for
NEC D4	NEC OD 9470 GTE D4	2 mm	Telecoms systems	NTT (Japan) Telecom Australia
SMA (US version)	Amphenol 905/6 Diamond GFS24A Hirschmann OSMA Radiall SFO SEL OS-5 Stratos 630	3·174 mm	Industrial systems LANs	Mil STD 1863 (NATO)
Stratos 430	Holtek 38000	3·8 mm	Telecoms systems	British Telecom
Diamond GFS 1-3 MMS 1/3	HFS 1/3 HMS 0/1	3·5 mm	Laboratory use	
F & G 2000	DIN 47 255	2·5 mm	Telecoms systems	Deutsche Bundespost
F Type NTT	Seiko SAP 1/2 Fujikura FFC OFTI 152 FC Hirose HRFC	2·5 mm	Telecoms systems	NTT Japan Finnish PTT Austria Telecom Sweden

Diamond GFS 11/13 MMS 11/13	HFS 11/13 HMS 10/11 Interoptics	2·5 mm	Laboratory use	W & G
Ericssons TSR 301		3·174 mm	Telecoms systems	Swedish PTT
Stratos 530		2·5 mm	Telecoms systems	Austrian PTT
Western Electric 1006 A (AT&T)	Dorran Photonics Sirti FCA OFTI 152/155 Auto Connector Runge	bi-conical	Telecoms systems	Italian PTT
Socapex	Souriau 8016	V-form	Telecoms systems	French PTT
Radiall PFO		optaball	Telecoms systems	
ANT A1		4 mm	Mobile OWG systems	
Siemens C42334	DIN 47 255	2·5 mm	Telecoms systems	

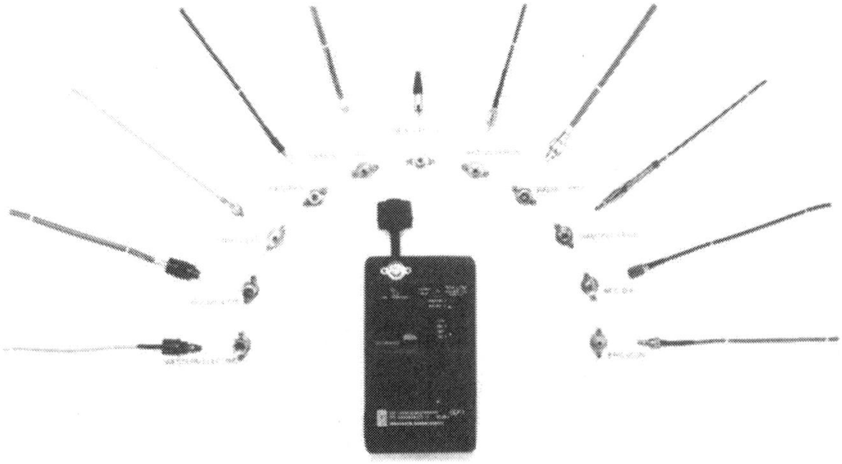

Plate 8.8
Showing some of the many connectors available for optical fibre systems (courtesy Wandel & Goltermann)

different standards and different, independent, development programmes. Table 8.1 and Plate 8.8 vividly illustrate the extent of the problem. An urgent requirement therefore exists for some form of international standardisation to bring about a reduction in the number of different types of pin construction, cylindrical diameter screw thread and length of fixing mechanism.

As the need for both an extremely simple and highly sophisticated apparatus already exists, the market for optical fibre measuring instruments is indeed a very fertile one and it is little wonder that manufacturers are almost falling over themselves to meet the current requirement of producing reasonably priced instruments to the customer.

This is certainly an area in the test equipment market that is unlikely to stagnate in the years to come!

9
Conclusion – the future

The use of optical fibres in the commercial world is split into three separate classes: **telecommunications** (mainly concentrating on high-performance trunk lines and area networks); **military and industrial** (where the emphasis is centred on digital communications and safety/security); and local area networks (LANs) or integrated services digital networks (ISDNs), which are mainly used in the business world.

More than 50 per cent of today's optical fibre applications are in the telecommunications field and according to a recent report by the *Kessler Marketing Intelligence* [18], the American optical fibre market is now expected to increase by an average annual growth rate of 11 per cent until 1992, despite a sluggish performance in 1986.

Confirming these expectations, the *UK Advisory Council on Science and Technology* (ACOST) in a recent report on optoelectronics, said that this field is of strategic importance and estimated that the world market would exceed £7 billion ($13 billion) annually.

ITT Corning of America, one of the first companies to become involved in optoelectronics, has estimated that the US market alone will require 2.3 million fibre km in 1989. There is, however, currently a slump in the market which is mainly due to the falling demands for optical fibres in long-haul and inter-office telephone links.

Interest in emerging applications such as government systems, oceanographic projects and industrial processor controlled networks are nevertheless rapidly growing. This is particularly relevant with respect to customers' premises and the use of LANs and ISDNs which are expected to increase by an annual average growth of 70 per cent.

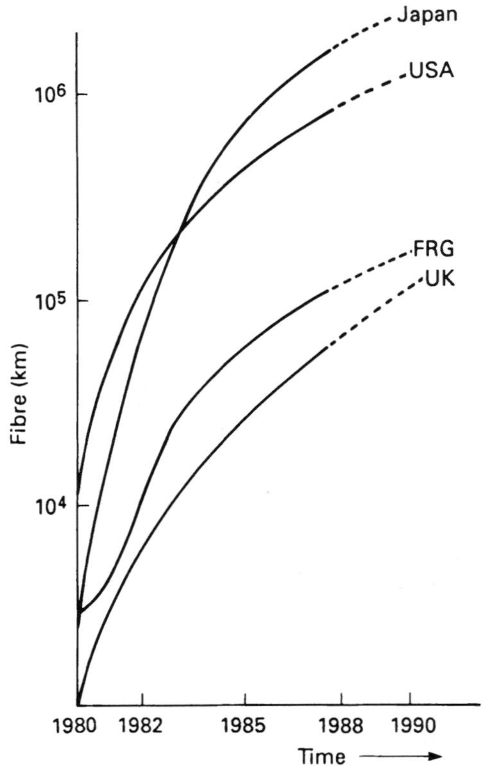

Figure 9.1 Predicted demand, in fibre km, for optical fibre systems as cumulative curves (information from Kessler Marketing Intelligence)

The predicted demand, in fibre km, for optical fibre systems is illustrated in Figure 9.1.

In Europe, the bulk of the optical fibre market is in France, West Germany and the United Kingdom:

- **France**; traditionally strong in telecommunication technology, became involved in optical fibres at a comparatively early stage – the **Biarritz installation** was the first in the world to bring services to the subscriber via optical fibres. France however has not, as yet, adopted optical fibres for trunk networks
- **Germany**; has also been fairly conservative in adopting optical fibres and although experimenting with an integrated glass fibre network in a number of cities (see description of BIGFON in Chapter 7) and recently com-

mitted to completing all of its future trunk installations with single-mode fibres, the initial impetus has waned

- **United Kingdom**; extensive networks of single-mode trunking have already been installed for telecommunications and data and much more is planned for the future together with a strong commitment to LANs and ISDNs. The Advisory Council on Science and Technology (ACOST) has suggested that the British Government should set up a stand-alone optical fibre network between its establishments. This, they argue would offer far more security than existing data networks and provide many more services including telex, facsimile, data transfer, high fidelity sound, television library and information.

Other European markets include: Switzerland (currently installing single-mode trunking between cities as part of their ISDN network); Denmark (replacing existing inter-island trunks with high capacity optical fibre cables; Sweden (which leads the world in telephone density – number of phones per capita – already uses single-mode fibres throughout and has plans to complete a nation-wide optical fibre long-haul network by 1990).

With such a world-wide interest and dependence in optoelectronics it is not surprising that the advance of this particular technology is progressing far more quickly than any other, and many new techniques are being developed.

Development trends

Advanced semi-semiconductor manufacture

With the use of non-gravitational environments (eg, on space stations), thin-film technology and advanced semiconductor manufacturing techniques holds great promise. For example:

Super lattice

Where molecular beam epitaxi puts down a layer which is only a few molecules thick.

Ballistic transport

Ballistic electron injection or ballistics, involves sending electrons across the discontinuous band gap created by layers of AlGaAs and GaAs. As the AlGaAs layer possesses a larger band gap, the resulting discontinuity provides a step potential that is picked up by the AlGaAs electrons as **kinetic energy** and electrons travel at speeds up to six times faster than they can in conventional structures. Use of microthin layers of AlGaAs and GaAs stabilises the path and thereby increases the chance of an electron hitting an atom (or positive ion) in the lattice and interacting with it.

Fluoride glasses

In May 1987 Corning Glass Works signed a joint research agreement with Nippon Telegraph and Telephone Company, to conduct basic research in **fluoride glasses**. At the International Symposium on Halide Glasses held in Shizuoka, Japan during May and June 1988, Corning and NTT announced that their research into fluoride glass had produced a 'sound fundamental and good first step' towards bulk-melt fluoride glasses.

The team reported that oxygen is instrumental in stabilising the inherently unstable fluoride glass and that fluoride glasses are believed to hold much promise as materials for future optical fibres. Theoretically, losses in fluoride glasses are two orders of magnitude less than other glasses, making the compositions desirable for low-loss, longer distance transmissions.

Another promising new approach being researched by the Corning/NTT team is vapour phase deposition of fluoride glass which can give high purities and produce preforms from which low-loss silica fibres can be drawn. So far [13], vapour deposited beryllium fluoride glass shows very low scattering losses. Corning has also been awarded a contract by the Navel Research Laboratory in Washington DC, for research into the beryllium fluoride for optical fibres [37].

It is anticipated that halide glasses will be an ideal material for future optical fibres and these are expected to be an enormous improvement over silica glass [18]; nevertheless much work remains.

High electron mobility ioniser

This has now become a reality and consists of an InGaAs layer upon a GaAs strip which is capable of producing bandwidths up to 250 GHz.

Re-emergence of multimode fibres

Manufacturers and users alike are now predicting the re-emergence of multimode fibres particularly in respect of military applications.

Moulded-clad single-mode fibres

Corning Glass Works, in August 1987, produced a new single-mode fibre that has a matched-clad, step-index design which offers greater resistance to microbending losses and gives improved performance. Attenuation losses are given as less than $0.5\,\mathrm{dB\,km^{-1}}$ at both 1300 nm and 1550 nm and it is reported that this fibre will also be available for wavelengths greater than these.

According to Corning [13], single-mode optical fibre has proved to be the most cost effective medium for transmitting at bit rates greater than $6\,\mathrm{Mbit\,s^{-1}}$ and at distances greater than 3 km.

This current trend towards specialised fibres reflects a growing user sophistication. Just as no one electrical cable can handle all electrical transmission needs, no single optical fibre is optimised for all optical transmission requirements.

Rare earth-doped fibre lasers and amplifiers

The incorporation of rare earth dopants into the core of single-mode optical fibres [38] has led to a new class of both active and passive fibre devices. These include: in-line optical amplifiers; absorption filters; distributed sensors; and, although still very much at the research stage, it is now possible even to fabricate an all-laser amplifier.

Rare earth-doped single-mode fibre lasers have been shown to be efficient, compact and versatile sources of laser radiation and it is anticipated that these devices will be used in a wide variety of systems applications in the near future.

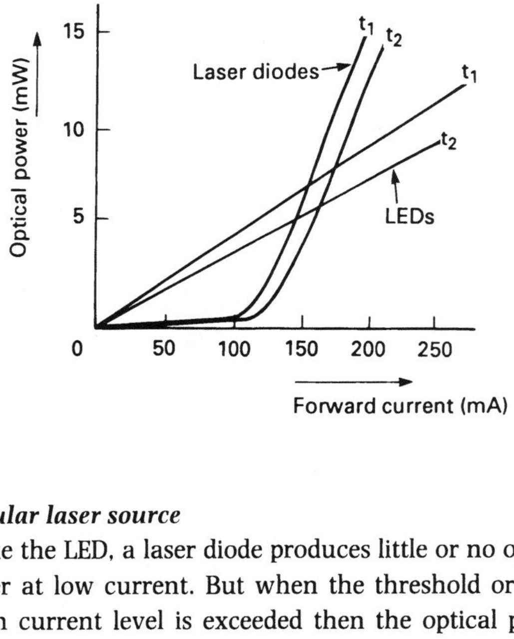

Figure 9.2 Optical power against forward current characteristics of typical lasers and LEDs

Modular laser source

Unlike the LED, a laser diode produces little or no optical power at low current. But when the threshold or minimum current level is exceeded then the optical power rises dramatically with a comparatively small increase in current. This is illustrated in the characteristics shown in Figure 9.2. Although an LED loses very little optical power as temperature increases (approximately 0.7 per cent degree [39]), a laser diode is very susceptible to temperature changes and an increase of only a few degrees can cause the laser to cease its lasting operation. Because of this, heat extraction, together with temperature stabilisation is very important.

Another point to be remembered is the established fact that an LED is capable of producing a far greater optical power density (eg, $4\,W\,cm^{-2}$) than a laser (typically $1\,W\,cm^{-2}$), but to ensure that the maximum amount of power is transferred to the fibre it is essential that laser diodes are protected against environmental ageing effects such as dust or dirt, organic vapors and humidity etc.

In recent years, encapsulating the laser diode into a hermetically-sealed, temperature-controlled, loss-free enclosure or module that is environmentally matched to the fibre has effectively overcome these problems and it has also produced an excellent method for ensuring that the junction losses between the laser source and optical fibre are minimised.

A typical example of such an **encapsulated laser diode** is shown as a block diagram in Figure 9.3a, and as cutaway in Figure 9.3b.The internal temperature of the module is measured by a thermistor (attached to the laser's heatsink) which produces the control for a **Peltier cooler** which ensures that the laser diode is held to an almost constant temperature.

To minimise matching losses, the junction at the interface of the laser diode and the optical fibre is usually completed inside the module and an external pigtail made available for connection to the optical waveguide.

A photocurrent is produced by a monitor diode using the light from the rear mirror and this is fed to an electrical control and drive circuit to provide an external, variable, optical power control.

Development trends in higher-order optical fibre transmission systems

In modern system designs a transmission rate of $8\,Mbit\,s^{-1}$ is normally used for the first optical transmission window (840 to 880 nm) and, occasionally, at

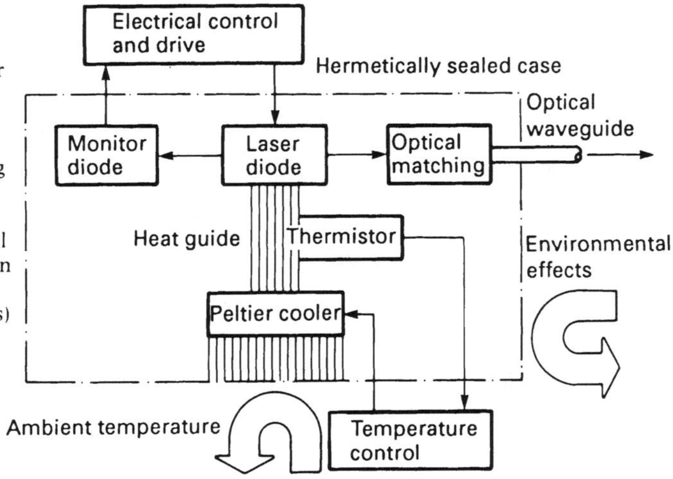

(a)

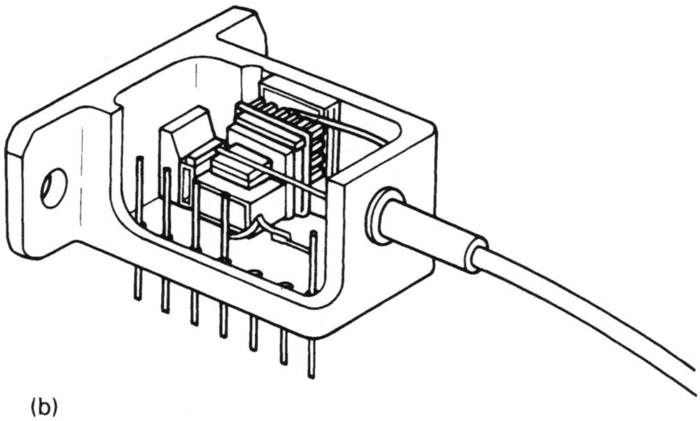

(b)

Figure 9.3
A typical encapsulated laser diode (a) a block diagram of operation (b) cutaway, showing monitor diode, laser diode, thermistor, as well as Peltier cooler (in background) (courtesy Siemens)

the second optical wavelength (1300 nm). In order to achieve optimum conditions, a laser diode is normally used as the transmitter for an 8 Mbit s^{-1} system while the receiver is usually an avalanche photodiode. These 8 Mbit s^{-1} systems would normally be used for such applications as branch carriers for digital exchanges, video telephony, data etc, and, as high transmission bandwidths are required, graded-index or single-mode fibres are invariably used.

In the 1970s, the possibility of increasing the transmission bandwidth by operating a system at a rate of 34 Mbit s^{-1} was actively researched and in 1979 the Deutsche Bundespost successfully opened a 15 km link between Frankfurt-Ginnheim and Oberursel [40].

The larger the bandwidth, the lower the losses a fibre may have and, while graded-index fibres are perfectly acceptable for 8, 34 and possibly 140 Mbit s^{-1} transmission rates, when this is increased to 560 Mbit s^{-1} single-mode fibres (operating at wavelengths of 1300 nm or so) have to be used.

Studies are now progressing to develop systems capable of operating in the Gbit s^{-1} range and at wavelengths in excess of 1500 nm.

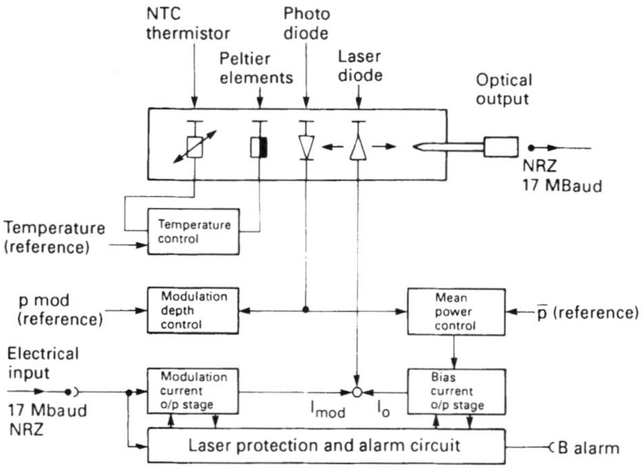

Figure 9.4 Functional diagram of an optical transmitter

During the past few years there has also been a number of radical changes to the transmission system and these have resulted in vast improvements to both **transmitters** and **receivers**. For example, the 8 Mbit s^{-1} system block diagram shown in Figure 9.4 uses the multimode emission spectrum of an GaAlAs oxide stripe laser diode. To ensure stable laser operation, a Peltier cooler and temperature sensor are incorporated into the hermetically-sealed laser module.

Also contained in the laser module is a drive circuit

which is used to regenerate the amplitude of the signal and supply the modulation current for the laser diode. Biasing, to permit the light pulse rise-time being delayed with respect to the drive current pulse is also used.

Regulation of the modulation depth and optical power is achieved with control loops and an alarm circuit is incorporated to ensure the laser diode is not overloaded.

The receiver, on the other hand, normally uses an avalanche photodiode to convert the received optical signal into an electrical signal. The current pulses of the photodiode are converted into voltage pulses by a gain-controlled pre-amplifier called a **transimpedence amplifier.** This is followed by a dual-gated field effect transistor (FET) amplifier. A low-pass filter provides band-limiting. Such an optical receiver is shown in block diagram form in Figure 9.5.

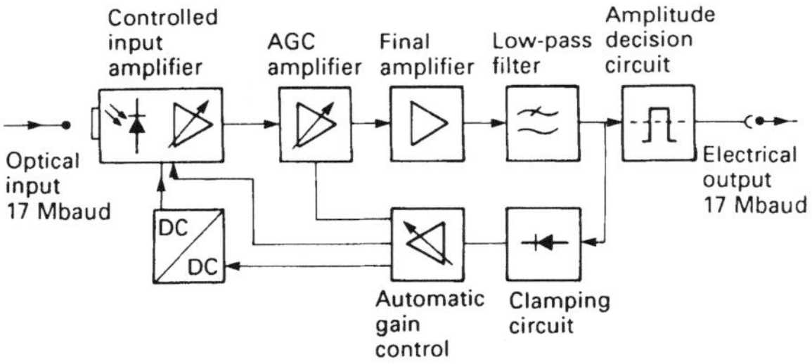

Figure 9.5 Functional diagram of an optical receiver

It is not uncommon to find PIN photodiode, FET and bipolar transistors integrated on a ceramic substrate. Normally, germanium avalanche photodiodes are used for wavelengths between 840 to 880 nm, while PIN photodiode with GaAs FETs are normally used for 1300 nm.

Lasers

With the introduction of the **distributed feedback laser** (DFB) which is now available for generation at 2.4 Gbit s^{-1}, even more advances are being made in laser technology and PIN-FET receivers capable of receiving such a high data rate are also available. These devices will provide substantial increases in the amount of individual traffic channels that can be sent over an optical fibre as well as enabling the spacing between regenerators to be increased [41]. Commercial development of all-optical amplifiers promises considerable savings especially in the cost of optical communication links.

Industry

The increased use of microelectronics in industrial controls has resulted in extremely high signalling rates being present in the system. The problem, of course, with high signalling rates is the requirement for low signalling voltages: this is one of the main reasons why industry is turning to optical fibres and a few of these applications are given here.

Leak detection

Whiteman Controls Corporation of America has designed and installed a leak detection method for double-walled piping systems containing hazardous fluids. The theory behind this is very simple but extremely effective and consists of a number of minute pressure sensors which are placed at regular intervals along the pipe. The sensors are joined to an electro-optical connector and the light is passed via the optical fibre cable to an optoelectronic interface located some 2 to 3 km away from the hazardous area.

Optical fibre drill bit temperature monitor

An optical fibre drill bit temperature monitor has been designed by Vanzetti Systems for use with printed circuit board laminate drill bits. An optical fibre infra-red detector head, connected to an optical fibre cable looks directly at and measures the temperature of the drill bit. The infra-red head is capable of detecting overheating of drill bits (due to dulling or overuse) which is then passed as a signal to an alarm circuit via the fibre cable.

Optical fibre soldering iron

Although still very much at the drawing board stage, the development of an optical fibre soldering iron, that can be used for field maintenance as well as in the laboratory, is well advanced.

Optical computing

The fastest computer in the world is the recently-developed (November 1987) BBC Andromeda, but even this has to rely on step processing or sequential functions and this severely restricts its capability. Light can emit hundreds of modes in parallel without interference and American SDI research is now investigating the possibility of an optical switch. This is effectively the first step towards optical computing which can be achieved by optically multiplexing many signals together and then switching one or any number of information channels in parallel.

Military

In the early 1960s, optical fibre technology, along with semiconductor and laser technology, became a particularly attractive possibility for strategic and tactical military applications. The military has always been concerned with radiation of energy from the transmission media and susceptibility to high energy electromagnetic radiation from nuclear weapons. Unfortunately, cost and doubtful performance factors have limited large-scale military implementation. With the emergence of more cost-effective manufacturing techniques, together with increased reliability, the military is now actively upgrading existing copper systems with optical fibre cables and extending digital ISDNs and LANs for fixed installations.

In keeping with this general trend, the American forces are now replacing a microwave system with a single-mode optical fibre backbone transmission cable in the Republic of Korea.

Conclusion

Often optoelectronics is depicted, especially by those who are afraid of it as a competing technology, as a science in its infancy, with inevitable teething problems. As the costs of cables and connectors are reduced, however, the non electrical nature of fibres will result in optoelectronics becoming more and more popular.

During the 1980s the use of optical fibres has increased considerably in both commercial and military environments. Without doubt, optoelectronics will become the dominating technology of the 1990s.

References

1 Aschoff B., *Optische Nachrichtenubertragung un Klassischen Alterum*, Nachricktentechn Z (NTZ) 30, pp. 23–8, 1977.

2 Harfield Alan G., *Blandford and the Military*.

3 Deacon Richard, *The Silent War*, David & Charles, 1978.

4 Steinbuch K., *Die Informierte Gesellschaft*, Stuttgart, pp. 60–6, 1966.

5 *The Illustrated London News*, p. 426, 8 November 1879.

6 Archer J. D., *Manual of Fibre Optic Communication*, STC Components, p. 4, January 1984.

7 de Vries Leonard, *Victorian Inventions*, John Murray, 1971.

8 *Fibre Optics Handbook*, Hewlet Packard GmBh FRG, October 1983.

9 Kostler, Gunter, *Fibre Optic Projects for the Deutsche Bundespost and Other Users in Europe*, 1983

10 Dawson J. B., *Fibre Optic Communications*, School of Signals Blandford, 1986.

11 Geckler Seigfried, *Physical Principles of Optical Waveguides*, 1983.

12 Archer J. D., *Manual of Fibre Optic Communication*, January 1984.

13 Keck Donald and Vandenoestine Robat, 'Updating the art', *Telephony*, pp. 48–57, 28 March, 1988.

14 Comité Consultatif International de Telegraphie et Telephonie.

15 Horiguchi M., Spectral Losses of Low OH Content Optical Fibres, *Electrol Lett*, pp. 310–11, 1976.

16 Gier Jurgen and Panzer Klaus, *Principles of Optoelectronic Signal Transmission*, 1983.

17 Braun Ewald and Keil Heinrich, *Optical Waveguide Transmission Systems*, 1983.

18 *Fibreoptic Product News*, Vol. 2, No. 5, May 1987.

19 Goldmann Horst, *Installation of Fibre Optic Cables*, 1983.

20 Clemen Christian, Heinen Jochen and Plihal Manfred, *High Radiance Light Emitting Diodes for Optical Transmitters*, 1983.

21 Volkel G. and Wallner E., *Quality Assurance Methods for Optical Semiconductors*, 1983.

22 Wolf H. D., Mettler K and Zschauer K., High Performance 880 nm (GaAl) As/GaAs Oxide Stripe Lasers with low Degradation Rates at Temperatures up to 120 degs C, *Jpn Journal of Applied Physics*, Vol. 2, pp. L693–6.

23 Ammam Marcus Christian, Mettler Klaus and Wolf Hans Dietrich, *Laser Diodes – High Power Light Sources for Optical Communications*, 1983.

24 Weyrick, Claus and Zschauer, Karl Heinz, *Principles of Optoelectronic Conversion*, 1983.

25 Tsuchiya T. and Imolo N., Dispersion Free Single Mode Fibre in 1,5 μm Wavelength Region, *Electrol Lett*, pp. 476–8, 1979.

26 Cooper R., Connector Splice Loss Measurements: the Laboratory versus the Real World, *IFOC*, Vol. 3, No.3/4, pp. 38–46, 1982.

27 Witt Gerd, *Closures and Containers for Fibre Optic Cables*, 1983.

28 *New Electronics*, p. 14, 12 May 1987.

29 Dufar Ian, *Flexible Access Systems*, British Telecom, 1988.

30 *Transmission Technologies – Fibre Optics*, Communications Systems Worldwide, p. 16, March 1987.

31 *Siemens Telecom Report*, 'Special Issue; Optical Communications', Siemens AG Munchen, 1983.

32 Bark P., Szenlesi R. and Istvan O., *Fibre Optic Projects in the USA*, 1983.

33 Braun Erward, *Bigfon Heralds the Start of a New Era in Telecommunications*, pp. 189–95, 1982.

34 Braun Erward, *Bigfon Points The Way to New Forms of Communication*, pp. 10–14, 1983.

35 British Telecom.

36 Rocks, M., Optimierung von Digitalen Mehrstufen Lichtleiterubertragungs System, *Frequenz 33*, pp. 290–8.

37 Signalgram, *Signal*, p. 247, June 1988.

38 Payne D. N. and Reekie L., *Overview of Rare Earth Doped Fibre Lasers and Amplifiers*, 1988.

39 Grau, G., *Optische Nachrichtentechnik Berlin*, Heidelberg, New York Springer Verlag, 1981.

40 Muller K., Unterlass W. and Weber W., *Two Years Experience With the First Operational Fibre Optic Route in the Deutsche Bundespost Network*, pp. 246–9, 1981.

41 *Electronic Technology*, STC Defence Systems New Devices For Fibre Optic Communications, vol. 20, pp. 3, January 1987.

Index

Printed in France by Amazon
Brétigny-sur-Orge, FR

51786488R00098